RSC Paperbacks

A WORKING METHOD APPROACH FOR INTRODUCTORY PHYSICAL CHEMISTRY CALCULATIONS

Numerical and Graphical Problem Solving

BRIAN MURPHY[1], CLAIR MURPHY[2] AND BRIAN HATHAWAY[2]

[1]*Department of Chemistry*
University of Wales, Cardiff, UK,

and

[2]*Department of Chemistry,*
University College Cork, Ireland

ISBN 0-85404-553-8

A catalogue record for this book is available from the British Library

Published by The Royal Society of Chemistry, Thomas Graham House, Science Park, Milton Road, Cambridge CB4 4WF, UK

Typeset by Computape (Pickering) Ltd, Pickering, North Yorkshire, UK
Printed by Athenaeum Press Ltd, Gateshead, Tyne and Wear, UK

A WORKING METHOD APPROACH FOR INTRODUCTORY PHYSICAL CHEMISTRY CALCULATIONS

Numerical and Graphical Problem Solving

RSC Paperbacks

RSC Paperbacks are a series of inexpensive texts suitable for teachers and students and give a clear, readable introduction to selected topics in chemistry. They should also appeal to the general chemist. For further information on available titles contact:

Sales and Promotion Department
The Royal Society of Chemistry
Thomas Graham House
Science Park, Milton Road
Cambridge CB4 4WF, UK

New Titles Available

Archaeological Chemistry
by A. M. Pollard and C. Heron
Food – The Chemistry of Its Components (Third Edition)
by T. P. Coultate
The Chemistry of Paper
by J. C. Roberts
Introduction to Glass Science and Technology
by James E. Shelby
Food Flavours: Biology and Chemistry
by Carolyn L. Fisher and Thomas R. Scott
Adhesion Science
by J. Comyn
The Chemistry of Polymers (Second Edition)
by John W. Nicholson
A Working Method Approach for Introductory Physical Chemistry Calculations
by Brian Murphy, Clair Murphy and Brian Hathaway

How to Obtain RSC Paperbacks

Existing titles may be obtained from the address below. Future titles may be obtained immediately on publication by placing a standing order for RSC Paperbacks. All orders should be addressed to:

The Royal Society of Chemistry
Turpin Distribution Services Ltd.
Blackhorse Road
Letchworth
Herts: SG6 1HN, UK
Telephone: +44 (0) 1462 672555
Fax: +44 (0) 1462 480947

Preface

The basis of physical chemistry is the ability to solve numerical problems. It is generally agreed that it is not with inorganic and organic chemistry that first-year and preliminary-year undergraduate students have the greatest difficulty, but instead the numerical problem-solving aspect of physical chemistry. The global trend of below-average marks in physical chemistry in first-year and preliminary-year chemistry papers needs to be addressed.

The preparation of textbooks has been made much easier by the improvements in the technology of book production. This has resulted in the production of much more colourfully attractive textbooks of general and introductory chemistry. This would not be a problem if the basic principles of chemistry were still clearly identifiable. Unfortunately, often this is not the case and the principles, even when well described, are lost beneath a wealth of factually unconnected data, that is unnecessary for the student to learn.

This is particularly apparent in the sections on basic introductory physical chemistry. Although many of these 1000-page textbooks contain well-written individual chapters on thermochemistry, equilibrium, electrochemistry and kinetics, with attractive diagrams, the fundamentals are sometimes lost in a sea of historical facts. No connectivity between the chapters is introduced and the impression is that each subject is divorced from the other sections. What is more disturbing is that although numerical problems and solutions do appear in such textbooks, no logical stepwise procedure is presented, leaving the student totally isolated when faced with a similar problem. Equally, the appearance of 60–70 numerical problems at the end of each chapter is unrealistic and inappropriate at this level. This approach is fine in more advanced physical chemistry textbooks, but such complexity and number of problems *is not suitable* for a basic undergraduate introductory course on physical chemistry!

The ability to solve numerical problems is the foundation of any physical chemistry course. This text-book is dedicated to those first-year and preliminary-year chemistry students who have not previously taken A-level or Leaving Certificate chemistry, or any student who finds major difficulty with physical chemistry problems at this level. Sadly, the skill of numerical problem-solving is being relegated to a bare nothing in secondary-level chemistry courses. The amount of physical chemistry that is taught is progressively being eroded away in schools, leaving first-year and preliminary-year chemistry students facing a complete cultural shock when posed with the volume of numerical problems which they must solve, on entrance to universities and tertiary-level institutions. This worrying trend may account for the poor first-year and preliminary-year examination results returned in physical chemistry papers.

Current textbooks do little to identify and confront this problem. A gentle systematic step-wise approach of a 'Working Method' is one method of addressing this decline in marks and the difficulty that students face with physical chemistry. Ironically, in many textbooks, working methods are mentioned; however, these appear with great infrequency. The aim of this textbook is to treat each numerical problem in introductory physical chemistry with the systematic step-wise approach of the 'working method'. Students need to gain confidence in tackling numerical problems and a systematic approach can certainly help. *The text does not imply offering 'recipes' for solving such problems, but hopefully it will encourage students to think out their own approach.* We believe that if students can get started on such problems, this approach will encourage the students to think for themselves when faced with more challenging problems. Equally, a working method on graphical problems is presented, a rarity in physical chemistry textbooks at this level, and yet essential in so many other subjects in science, such as physics, biochemistry, nutrition, geography, optometry, food science, *etc.*; undergraduate students of these courses rely on a good basic chemistry course to develop such a skill.

The present text tries to overcome the limitations of these texts, by covering the basic principles of introductory physical chemistry in a short and concise way. Each chapter contains an introduction, followed by a typical examination question at the appropriate level, and then takes the student stepwise through the working method.

This text is written against a series of computer-aided learning (CAL) tutorials. These tutorials have been in use for the past four years at University College Cork (UCC) (and more recently at the University of Wales, Cardiff and Dublin City University), and are extremely popular with the 300 students per year taking the course. With such a back-up to the courseware, the main recommended physical chemistry textbook, the large and small group tutorials and the lecture course, the student should not feel so isolated with the problems associated with physical chemistry.

The use of the CAL courseware in UCC is entirely optional and supplementary to the normal teaching programme (namely, lectures, practicals, large and small group tutorials), but the interactive nature of the courseware, especially for numerical problem solving, is very attractive to the students, particularly to those with a weak chemistry background. As the courseware is based upon UCC-type examination questions and also reflects the UCC lecturer's approach to his teaching, the tutorials are not 100% transferable to other third-level institutions, but the physical chemistry tutorials are available from the authors to illustrate the approach taken in writing the courseware and are available free of charge on the Internet at URL:http://www.cf.ac.uk/uwcc/chemy/murphybm/bm1.html. This generally follows the approach of the individual chapters in the present text and, in any case, the authors firmly believe that the *best* courseware should be written in-house, to best reflect the approach of the course lecturer involved.

One final point. This text is based on the current first-year science chemistry course of a four-year B.Sc. degree course taken at University College Cork. The majority of students taking this course are non-intending chemists. Although the text covers most of the main areas of a typical general chemistry course, the authors *do not claim* in any way that this material is the most appropriate for such a course; indeed, many universities may include topics such as spectroscopy in such a course, and may prefer to change the order of the subjects taught. This, however, is *not* the purpose of this text. For example, the two chapters on kinetics are relegated to the end of the text, as the authors have found that students have trouble with some of the maths in this section. Also, the two chapters on electrochemistry are slightly expanded, since many students have expressed concern over the presentation of such material in other texts. So, in conclusion, this text is in many senses in response to the needs of the non-intending chemistry students who have struggled for far too long in physical chemistry at this level. However, it is hoped that lecturers and teachers will also find this a useful text, and will use this text as a basis for encouraging students to think out problems for themselves, before going on to more challenging problems!

Note: while every effort has been made to eliminate errors in this book, inevitably some will have crept through. The authors would appreciate notification by readers of any errors. They may be contacted as follows:

Brian Hathaway: Tel. 353–21–902379
Fax 353–21–274097
e-mail stch@bureau.ucc.ie

Brian Murphy: Tel. 353–1–7045000
Fax 353–1–7045503
e-mail murphybr@ccmail.dcu.ie

Brian Murphy and Brian Hathaway

Contents

Acknowledgements

The authors wish to acknowledge Dr Ross Stickland, Physical Chemistry Section, Department of Chemistry, University of Wales, Cardiff and Gillian Murphy, Department of Chemistry, University of Cork, for proof-reading this text. The following are also acknowledged for useful suggestions and continued support: Professor Wyn Roberts, Head of Department, University of Wales, Cardiff, Dr Gary Attard, Dr Eryl Owen, Physical Chemistry Section, and Professor Michael B. Hursthouse, Structural Chemistry, Department of Chemistry, University of Wales, Cardiff.

Chapter 1

Introduction to Physical Chemistry: Acids and Bases, The Gas Laws and Numerical and Graphical Problem Solving

INTRODUCTION

This chapter is a brief introduction to many of the assumptions made in the remainder of this text and the basis of physical chemistry type problems. The spider diagram of Figure 1.1 illustrates the various sections.

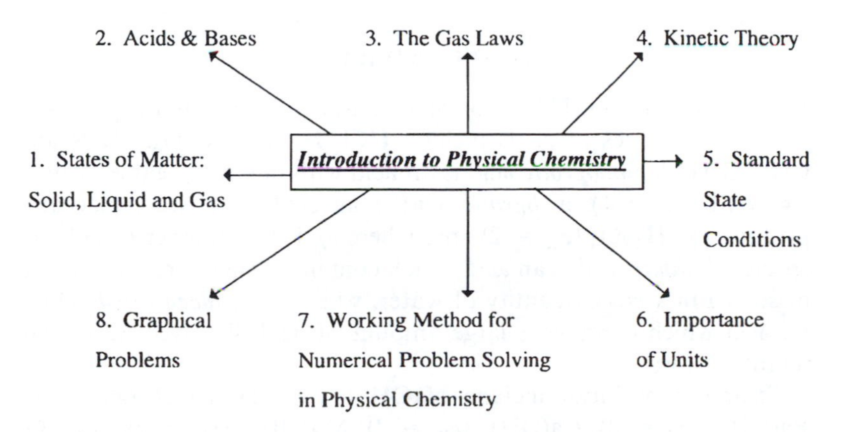

Figure 1.1 *Summary of the contents of Chapter 1*

STATES OF MATTER

Matter is the chemical term for materials. There are three states of matter: the solid phase (s), the liquid phase (l) and the gaseous phase (g). In the solid phase, all the atoms or molecules are arranged in a highly ordered manner [Figure 1.2(a)], whereas in the liquid phase [Figure 1.2(b)], this ordered structure is not as evident. In the gaseous phase [Figure 1.2(c)], all the particles are moving at high velocity, in random motion. The disorder or entropy, S, is at its maximum in the gaseous phase [Figure 1.2(c)].

(a) Solid (s)

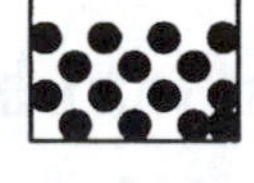

(b) Liquid (l)

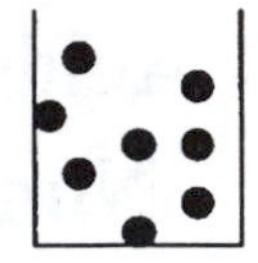

(c) Gas (g)

Increasing disorder or entropy, S

Figure 1.2 *States of matter—solid, liquid and gas*

If a species is dissolved in water, it is said to be in the aqueous phase (aq), and the symbol can be represented as a subscript, *e.g.* $HCl_{(aq)}$.

ACIDS AND BASES

An ***acid*** is a proton (H^+) donor and a base is a proton acceptor, *e.g.* (OH^-). Examples of acids include HCl, H_2SO_4, HNO_3, HCN and CH_3CO_2H. A ***monoprotic*** acid is an acid with one replaceable proton, *e.g.* HCl ($e_A = 1$); a ***diprotic*** acid is an acid with two replaceable protons *e.g.* H_2SO_4 ($e_A = 2$) *etc.*, where e_A is the number of reactive species. A ***dilute*** acid is an acid which contains a small amount of acid dissolved in a large quantity of water, whereas a ***concentrated*** acid is an acid which contains a large amount of acid dissolved in a small quantity of water.

Examples of bases include NaOH ($e_B = 1$), KOH ($e_B = 1$), $Ba(OH)_2$ ($e_B = 2$), $Ca(OH)_2$ ($e_B = 2$), $Mg(OH)_2$ ($e_B = 2$), Na_2CO_3 ($e_B = 2$), NH_3 and CH_3NH_2, where e_B is the number of reactive species. *e.g.* OH^-. An acid combines with a base to form a salt and water:

i.e.	ACID	+	BASE	→	SALT	+	WATER
e.g.	HNO_3	+	NaOH	→	$NaNO_3$	+	H_2O

In general, an acid can be represented as **HA**, where **HA** → $\mathbf{H^+}$ + A^- or, *more precisely*, **HA** + H_2O → A^- + H_3O^+, since all aqueous protons are solvated by water. Likewise, a base containing hydroxide anions, OH^-, can be represented as **MOH**, where **MOH** → M^+ + $\mathbf{OH^-}$.

When an acid donates a proton, H^+, it is said to form the ***conjugate base*** of the acid, *i.e.* HA (acid) ⇌ H^+ + A^- (conjugate base). The conjugate base is a base since it can accept a proton to reform HA, the acid. Similarly, when a base accepts a proton, H^+, the ***conjugate acid*** of the base is said to be formed, *i.e.* B (base) + H^+ ⇌ HB^+ (conjugate acid). The conjugate acid is an acid since it can donate a proton, H^+, and reform the base, *e.g.* NH_4^+ (conjugate acid) ⇌ NH_3 (base) + H^+.

Ions, Cations, Anions, Oxyanions and Oxyacids

Ions are charged species, *e.g.* Na^+, Cl^-, NH_4^+ *etc.* ***Cations*** are positively charged ions, *e.g.* Na^+, NH_4^+, Mg^{2+}, H_3O^+ *etc.* ***Anions*** are negatively charged species, *e.g.* OH^-, Cl^-, O^{2-} *etc.* A useful way of remembering this is the two n's, *i.e.* a**n**ion = **n**egatively charged ion! An ***oxyanion***, as its name suggests, is an anion containing oxygen, *e.g.* NO_3^-, SO_4^{2-} *etc.* An ***oxyacid*** is the corresponding acid of the oxyanion, *e.g.* HNO_3 and H_2SO_4 are the oxyacids of the nitrate and sulfate oxyanions respectively. The oxidation state or oxidation number of the nitrogen, N^V and the sulfur, S^{VI}, is the same in both the oxyacid and the oxyanion. Table 1.1 is a summary of some of the common oxyanions and their corresponding oxyacids.

Table 1.1 *Summary of some of the common oxyanions, the corresponding oxyacids, charges, oxidation numbers and the number of replaceable hydrogens*

Charge	*Oxyanion*	*Oxyacid*	e_A
−1	$N^{III}O_2^-$ (Nitrite)	$HN^{III}O_2$ (Nitrous acid)	1
−1	$N^{V}O_3^-$ (Nitrate)	$HN^{V}O_3$ (Nitric acid)	1
−2	$S^{IV}O_3^{2-}$ (Sulfite)	$H_2S^{IV}O_3$ (Sulfurous acid)	2
−2	$S^{VI}O_4^{2-}$ (Sulfate)	$H_2S^{VI}O_4$ (Sulfuric acid)	2
−3	$P^{III}O_3^{3-}$ (Phosphite)	$H_3P^{III}O_3$ (Phosphorous acid)	3
−3	$P^{V}O_4^{3-}$ (Phosphate)	$H_3P^{V}O_4$ (Phosphoric acid)	3
−1	$Cl^{V}O_3^-$ (Chlorate)	$HCl^{V}O_3$ (Chloric acid)	1
−1	$Cl^{VII}O_4^-$ (Perchlorate)	$HCl^{VII}O_4$ (Perchloric acid)	1
−2	$C^{IV}O_3^{2-}$ (Carbonate)	$H_2C^{IV}O_3$ (Carbonic acid)	2

THE GAS LAWS—IDEA OF PROPORTIONALITY

Boyle's Law

Pressure is defined as the force acting on a unit area, *i.e.* $p = F/A$. The unit of pressure is the newton per square metre, $N\ m^{-2}$ or the Pascal, Pa. At sea level, the pressure due to the weight of the earth's atmosphere is approximately 10^5 Pa. The bar, is often used as the unit of pressure in problems in physical chemistry, where 1 bar = $10^5\ N\ m^{-2}$ or 10^5 Pa.

Consider the effect of a piston pressing down on a fixed mass of gas of initial pressure $p_{initial}$ and initial volume $V_{initial}$, (Figure 1.3).

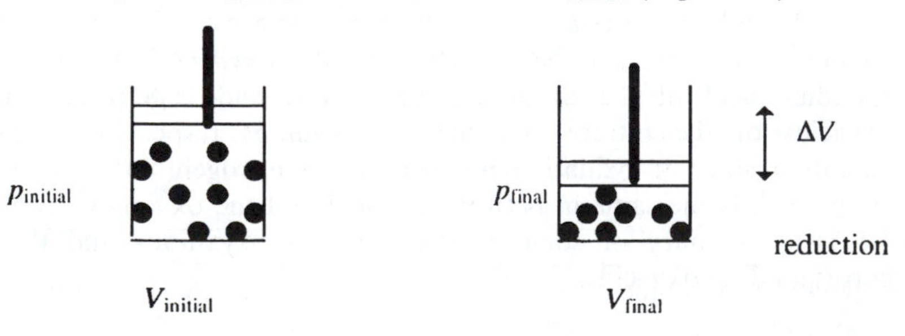

Figure 1.3 *Application of pressure on a definite mass of gas at constant temperature*

As the pressure applied by the piston is increased, the volume of the gas decreases (*i.e.* the space it occupies), if the temperature is kept constant. This is Boyle's Law – the volume of a definite mass of gas at constant temperature is inversely proportional to its pressure.

i.e. ***Boyle's Law:*** $V \propto 1/p$
or $V = k/p$, where k is a constant of proportionality.

Example: A sample of gas G used in an air conditioner has a volume of 350 dm^3 and a pressure of 85 kPa at 25 °C. Determine the pressure of the gas at the same temperature when the volume is 500 dm^3.

Solution:
The gas is at constant temperature, and therefore Boyle's Law can be applied:

Hence initially, $V_1 = k/p_1$ or $k = p_1 V_1$
Hence $k = (85 \text{ kPa}) \times (350 \text{ dm}^3)$
$= 29750 \text{ kPa dm}^3$
However finally, $p_2 = k/V_2$
$= 29750/500$
$= 59.5 \text{ kPa}$

Answer: Final Pressure = 59.5 kPa

Charles's Law

In contrast, if the pressure is kept constant, the volume of a definite mass of gas will increase, if the temperature is raised. This is Charles's Law: the volume of a definite mass of gas at constant pressure is directly proportional to its temperature.

i.e. ***Charles's Law:*** $V \propto T$
or $V = kT$, where k is a different constant of proportionality.

Example: A sample of gas G occupies 200 cm^3 at 288 K and 0.87 bar. Determine the volume the gas will occupy at 303 K and at the same pressure.

Solution:
The gas is at constant pressure, and therefore Charles's Law can be applied:

$$\text{Hence initially, } V_1 = kT_1 \quad \text{or} \quad k = V_1/T_1$$
$$\text{Hence} \quad k = (200 \text{ cm}^3)/(288 \text{ K}) = 0.69\dot{4} \text{ cm}^3 \text{ K}^{-1}$$
$$\text{Now } V_2 = kT_2, \text{ so,} \quad V_2 = (0.69\dot{4} \text{ cm}^3 \text{ K}^{-1}) \times (303 \text{ K}) = 210.42 \text{ cm}^3$$

Answer: Final Pressure = 210.42 cm³

Avogadro's Law

This states that equal volumes of gases, measured at the same temperature and pressure, contain equal numbers of molecules.

> *i.e.* ***Avogadro's Law:*** $V \propto n$, where n = the amount of gas (measured in moles).

The ***mole*** is defined as the amount of a substance which contains as many elementary species as there are atoms in 12 g of the carbon-12 isotope. A mole contains 6.02205×10^{23} particles, where N_A is defined as ***Avogadro's constant*** *i.e.* $N_A = 6.02265 \times 10^{23}$ mol^{-1}.

Ideal Gases

An ideal gas is a theoretical concept, a gas which obeys the gas laws perfectly. If the three gas laws are combined, the resulting equation is the equation of state of an ideal gas:

$$\text{(a) Boyle's Law}: V \propto 1/p$$
$$\text{(b) Charles's Law}: V \propto T$$
$$\text{(c) Avogadro's Law}: V \propto n$$
$$\text{(a), (b) and (c)} \Rightarrow V \propto (Tn)/p$$
$$\Rightarrow pV \propto nT$$
$$\Rightarrow pV = knT,$$

where k is another constant of proportionality called the Universal Gas Constant, R.

> ***Ideal Gas Equation:*** $pV = nRT$

where p = pressure of the gas (measured in bar); V = volume of the gas (measured in dm^3); n = amount of gas (measured in mol); T = temperature of the gas (measured in K); R = Universal Gas Constant = 0.08314 dm^3 bar K^{-1} mol^{-1} (or 8.314 J K^{-1} mol^{-1}). The above equation can also be modified to take into account changes in temperature, ΔT, changes in volume, ΔV, and changes in the coefficients of gaseous reagents, $\Delta \nu_g$ respectively, *i.e.* $pV = nR\Delta T$, $p\Delta V = nRT$ and $pV = \Delta \nu_g RT$. ν_g represents the coefficients in a chemical equation. For example, in the reaction $N_{2(g)} + 3H_{2(g)} \rightarrow 2NH_{3(g)}$, $\Delta \nu_g = \Sigma[\nu(\textit{Gaseous products})] - \Sigma[\nu(\textit{Gaseous reactants})] = [(2)] - [(1) + (3)] = -2$.

Molar Volume

1 mole of an ideal gas measured at 25 °C and 1 bar pressure occupies 24.8 dm^3 (where 1 dm^3 = 1000 cm^3).

i.e. 1 mole of an ideal gas at 25 °C and 1 bar pressure occupies 24.8 dm^3.

Example: Calculate the amount of gas in moles in 2000 cm^3 at 25 °C and 1 bar pressure.

Solution:
At 25 °C and 1 bar pressure, 1 mole of an ideal gas occupies 24.8 dm^3

	i.e.	24.8 dm^3	≡	1 mol
		1 dm^3	≡	(1/24.8) mol
2000 cm^3	=	2 dm^3	≡	(2/24.8) mol = 0.081 mol

Answer: 0.081 mol.

KINETIC THEORY OF GASES

Kinetic energy is the energy a body possesses by virtue of its motion. The molecules of gases travel at high velocity and hence have high kinetic energy. The kinetic theory of gases is used to explain the observed properties of gases, of which Brownian motion provides good evidence. ***Brownian motion*** is the irregular zigzag movement of very small particles suspended in a liquid or gas. If tobacco smoke is placed in a small cell, well-illuminated on a microscope stage, the tiny particles appear to be moving at random, as shown in Figure 1.2(c). This is due to the smaller invisible molecules of the air, colliding with the particles of the smoke. This is Brownian motion.

The kinetic theory of gases is a model proposed to account for the observed properties of gases. In order that the model is applicable, certain assumptions are made. For this reason, gases are categorised into two types: (a) ideal gases (as defined previously) and (b) real gases (defined as non-ideal gases).

Assumptions of the Kinetic Theory for an Ideal Gas

1. Gases consist of tiny molecules, which are so small and so far apart that the actual size of the molecules is negligible in comparison to the large distance between them.
2. The molecules are totally independent of each other, *i.e.* there are no attractive or repulsive forces between the molecules.
3. The molecules are in constant random motion. They collide with each other and with the walls or sides of the container, which changes the direction of linear motion.
4. For each elastic collision, there is no net loss of kinetic energy. However, there may be transfer of energy between the particles in such a collision.
5. The average kinetic energy, of all the molecules is assumed to be proportional to the absolute temperature T (measured in K) *i.e.* kinetic energy $\propto T$.

Validity of the Assumptions of the Kinetic Theory for an Ideal Gas

Assumption 1 This is largely true, since the compressibility of gases is very high. However, at high pressures, when a gas is highly compressed, it is not valid to state that the physical size of gas molecules is practically negligible compared with the distances between the molecules.

Assumption 2 This assumption is approximate, since gases diffuse or spread out to occupy all space available to them, *i.e.* there must be no appreciable binding force between the molecules. However, van der Waals forces exist (intermolecular forces of attraction and repulsion), and with polar molecules, other attractive forces exist.

Assumption 3 This assumption is valid, as shown by Brownian motion.

Assumptions 4 and 5 These assumptions are true, since in any elastic collision, kinetic energy is not lost.

Conclusion of the Kinetic Theory of Gases

The kinetic theory of gases is a good approximation, used to explain the behaviour of real gases. The theory has to be modified at very high pressures and in the presence of van der Waals forces, and for polar molecules where stronger intermolecular forces of attraction are involved.

THE GENERAL GAS EQUATION

Consider a gas at temperature T_1, pressure p_1 and volume V_1, and another gas at temperature T_2 and pressure p_2. The volume of the latter gas can be determined easily using the equation:

$$\frac{p_1 V_1}{T_1} = \frac{p_2 V_2}{T_2}$$

A useful way of remembering this is '**p**eas and **V**egetables go on the **T**able'! In fact, given any five of the above variables, the sixth can be evaluated using the equation. This will form the basis of the worked example at the end of this chapter.

Standard State

The ***standard state*** of a body is the most stable state of that body at 25 °C and 1 bar pressure (its symbol is °, *e.g.* E°, ΔH°, ΔS°, defined later in Chapters 2, 3 and 6 respectively).

Standard State—Most Stable State—S.S.—25 °C and 1 bar pressure.

IMPORTANCE OF UNITS IN PHYSICAL CHEMISTRY

In physical chemistry questions, the International system of units (SI) should be used. In any problem, one of the first steps is to convert all units to SI; note especially that ***temperature must always be given in kelvins***, *never degrees centigrade*:

Remember: $T\,(\text{K}) = T\,(^\circ\text{C}) + 273$

e.g. 25 °C is equal to 298 K since (25 + 273) K = 298 K. Table 1.2 lists (a) the basic SI units, (b) the derived SI units, and (c) some examples of non-SI units, which are commonly used in physical chemistry.

Table 1.2(a) *Basic SI units*

Physical quantity	*Name of unit*	*Symbol*
Length (l)	Metre	m
Mass (m)	Kilogram	kg
Time (t)	Second	s
Electric current (I)	Ampère	A
Temperature (T)	Kelvin	K
Amount of substance (n)	Mole	mol

Table 1.2(b) *Some derived SI units*

Physical quantity	*Name of unit*	*Symbol*
Area (A)	Square metre	m^2
Volume (V)	Cubic metre	m^3
Density (ρ)	Kilogram per cubic metre	$kg\ m^{-3}$
Force (F)	Newton (N)	$J\ m^{-1}$
Pressure (p)	Newton per square metre	$N\ m^{-2}$
Work (w)	Joule (J)	N m
Electric charge (Q)	Coulomb (C)	A s
Potential difference (V)	Volt (V)	$J\ A^{-1}\ s^{-1}$
Heat capacity (C)	Joule per kelvin	$J\ K^{-1}$
Specific heat capacity (c)	Joule per gram per kelvin	$J\ g^{-1}\ K^{-1}$

Table 1.2(c) *Examples of some non-SI units*

Physical quantity	*Name*	*SI equivalent*
Volume (V)	Litre (l)	$1\ l = 10^{-3}\ m^3$
Pressure (p)	Bar	$1\ bar = 10^5\ N\ m^{-2}$

A GENERAL WORKING METHOD TO SOLVE NUMERICAL PROBLEMS IN PHYSICAL CHEMISTRY

1. Read the question carefully—do not be put off by the sheer length or intricacy of a question. If you do not see the wood for the trees immediately—*don't worry —all will be revealed if you use a stepwise systematic approach!* Just ***break down the question***, step by step!
2. If a chemical equation is involved, identify all the ***species present***, along with their states, *i.e.* (s), (l), (aq) or (g). This is particularly relevant in thermodynamics, equilibrium and electrochemistry questions.

3. Write down the ***balanced chemical equation*** if necessary, including the stoichiometry factors, ν_A, ν_B, ν_C and ν_D respectively: $\nu_A\,A + \nu_B\,B \rightarrow \nu_C\,C + \nu_D\,D$.
4. Identify all the ***data*** given in the question, including any constants which are involved, *e.g.* $R = 8.314$ J K^{-1} mol^{-1}, 1 F = 96 500 C mol^{-1} (defined later), 1 mole of a gas behaving ideally at 25 °C and 1 bar pressure occupies 24.8 dm^3.
5. Convert all ***units*** to the one system, *i.e.* change °C to K, hours to seconds, *etc.* Watch out especially for (a) standard state conditions, *e.g.* $E°$, $\Delta H°$, $\Delta S°$ and $\Delta G°$ parameters: 25 °C (298 K) and 1 bar pressure, and; (b) $R = 0.08314$ dm^3 bar K^{-1} mol^{-1} (used when the pressure is expressed in bar) or more generally $R = 8.314$ J K^{-1} mol^{-1}. For example, suppose you were asked to calculate the volume occupied by 0.5 mol of $N_{2(g)}$ at 280 K and 0.93 bar, using the ideal gas equation, *i.e.* $pV = nRT$, and rearranging, $V = nRT/p$. Since p, the pressure of the gas, is given in bar, $R = 0.08314$ dm^3 bar K^{-1} mol^{-1} must be used, ***not*** $R = 8.314$ J K^{-1} mol^{-1}. Therefore, V = (0.5 mol) × (0.08314 dm^3 bar K^{-1} mol^{-1}) × (280 K)/0.93 bar = 12.52 dm^3 .
6. Identify the ***unknown*** in the question, *i.e.* what quantity is being looked for specifically? Do not be afraid to sketch ***a simple diagram*** if this identifies the problem for you!
7. Write down all relevant ***formulae***. The question will most likely involve just one or, less likely, more than one, of these equations.
8. ***Substitute*** the known data into the equations in step 7 above (this may help to identify the unknown) and ***solve*** for the unknown. Remember, it is easier to rearrange the equations *before* substituting the numerical values.
9. Write down your ***answer*** (*never have* 'x = □' as your final answer) and ensure that the appropriate units are given. Remember, logarithmic values are dimensionless, *i.e.* they have no units. For example, suppose that the concentration of H_3O^+ ions in a solution is 0.04 M. Then, the pH (defined in Chapter 5) can be calculated using the standard equation, pH $= -\log_{10}[H_3O^+] = -\log_{10}(0.04\ \text{M}) = 1.40$, *i.e.* although the units of concentration are mol dm^{-3}, there are no units for pH, since logs are involved.
10. Re-read the question and answer any ***riders*** to the question!

Armed with such an approach, ***most*** numerical problems in introductory physical chemistry can be broken down and tackled, no matter how difficult they may first appear.

Worked Example

Example: A gas G, occupying 2.5 dm^3 at 0 °C and 1 bar pressure, is transferred to a 5.5 dm^3 container, where the pressure is 0.25 bar. In order for the gas in the new vessel to attain this pressure, what must the temperature of the gas be?

Solution:

1. Read the question carefully—looks complicated. . . ? Just break it down!
2. No balanced chemical equation is involved, so step 2 can be skipped in this case.
3. One species involved—a gas $G_{(g)}$!
4. Identify the data in the question: $V_1 = 2.5$ dm^3; $T_1 = 0$ °C; $p_1 = 1$ bar; $V_2 = 5.5$ dm^3; $p_2 = 0.25$ bar.
5. Convert temperature to kelvins: $T_1 = (0 + 273)$ K $= 273$ K.
6. Unknown $= T_2$!
7. Pressure, volume and temperature suggest that the required equation is: $(p_1V_1)/T_1 = (p_2V_2)/T_2$, *i.e.* 'Peas and Vegetables go on the Table'!
8. Rearrange the equation before substituting the numerical values: $T_2 = (p_2/p_1)(V_2/V_1)(T_1)$. Solve for T_2, *i.e.* $T_2 =$ [0.25 bar/1 bar] × [5.5 dm^3/2.5 dm^3] × [273 K] $=$ 150.15 K.
9. ***Answer: Temperature is 150.15 K.*** Notice how the units cancelled each other conveniently in step 8. In this question the temperature could have been left in °C, but it is good practice always to convert temperature to the absolute temperature, measured in K.

WORKING METHOD FOR GRAPHICAL PROBLEMS

A corresponding working method can be applied to graphical problems in physical chemistry, adopting the same approach.

1. Read the question carefully.
2. Identify the tabulated data given to you; tables of data normally mean a graph has to be plotted. Remember, you might not necessarily be told this in a problem.

3. Convert all units to the SI system, *i.e.* T should be expressed in K *etc.*
4. Having identified the data, try to establish a linear correlation between the two sets of data—remember it is *not* simply a case of plotting one set of data points on the x-axis and the other set of data points on the y-axis. Identify the linear equation in question: $\boldsymbol{y = mx + c}$, where m is the gradient of the graph and c is the intercept, the point where the graph cuts the y-axis when $x = 0$.
5. Create a table of the appropriate data, taking special care of:
   (a) logs: did you use natural logs to the base e, for example, or $\log_{10}$?
   (b) did you convert direct values (x) to their reciprocal values ($1/x$)?
   (c) units (*e.g.* logs: dimensionless; $1/T$: K^{-1}, *etc.*).

   Add as many extra columns as required. Keep the tabulated data vertical (*i.e.* go down the page); this will ensure you do not run out of space!

x-axis/unit	y-axis/unit

6. Examine the table from step 5. From this, write down the maximum and minimum values of x, and also the maximum and minimum values of y:

   Maximum value of x = □; Minimum value of x = □;
   Maximum value of y = □; Minimum value of y = □.

   This determines an appropriate scale for the graph. At this point, you might want to return to step 5, and *'round off'* any numbers for plotting purposes. Add additional columns to the table if necessary.

7. Draw the graph (on *graph paper*!), and remember the following points.
   (a) Every graph must have a title.
   (b) Label the two axes.
   (c) Include the units on the axes, but remember, there are no units for logarithmic values.
   (d) Maximise the scale of the graph for accuracy.
   (e) Draw the best-fit line through the set of points. It is not essential that the line contains any of these experimental points.

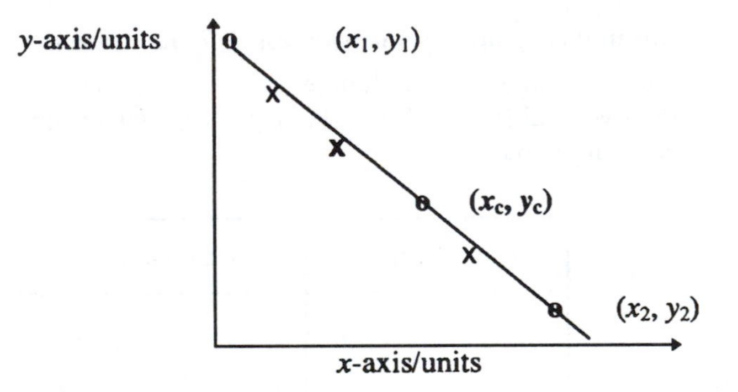

8. Determine the slope or gradient of the graph, by choosing two independent points on the line, at the two extremities, (x_1, y_1) and (x_2, y_2) respectively. Then apply

   $m = \Delta y/\Delta x = (y_2 - y_1)/(x_2 - x_1)$, and do not forget that the slope has units too.

9. The intercept of the graph, c, is then determined by examining where the graph cuts the y-axis (at $x = 0$). The units of c are obviously the y-axis units. If however, you find from the scale of your graph, that $x = 0$ is not included, c can still be determined, without extrapolating (extending) the graph. To do this, choose another independent point (x_c, y_c) on the line in the centre of the graph, and use the formula: $y = mx + c \Rightarrow y_c = mx_c + c \Rightarrow c = y_c - mx_c$, since m has already been determined in step 8.

10. From the values of m and c, determine the unknown parameter(s).
11. Answer any riders to the question.

In Chapter 9 on Chemical Kinetics, a worked example on graphical problems will be considered, using the above working method.

CONCLUSION

In Chapter 1, some of the basic concepts of physical chemistry are introduced, as well as the systematic step-by-step working methods which should be used where appropriate to tackle numerical or graphical problems in physical chemistry.

MULTIPLE-CHOICE TEST

1. Charles's Law states that:

   (a) $V = kn$ (b) $V \propto p$ (c) $V = kT$ (d) $pV = nRT$

2. Which of the following statements is incorrect?

   (a) The conjugate base of CH_3CO_2H is $CH_3CO_2^-$;
   (b) A concentrated acid is an acid which is dissociated completely in solution;
   (c) $H_2C_2O_4$, ethanedioic acid, is an example of a diprotic acid;
   (d) Avogadro's Law states that equal volumes of gases measured at the same temperature and pressure contain equal number of molecules.

3. The volume of a sample of carbon monoxide, CO, is 325 cm^3 at 15 bar and 520 K. What is the volume of the gas at 3.92 bar and 520 K?

   (a) 12.4 dm^3 (b) 0.00124 dm^3 (c) 1244 dm^3 (d) 1.24 dm^3

4. How many of the following statements are incorrect?

   * The pH of a 0.08 M solution of $HClO_4$ is 1.10 M.
   * Standard state conditions are 298 K and 1 bar pressure.
   * 1 bar $= 10^5$ Pa.
   * The kinetic theory of gases is valid at high pressure.

   (a) 1 (b) 2 (c) 3 (d) All 4

5. What is the density of $C_2H_{6(g)}$ at 175 kPa and 20 °C?
   (C, 12.01, H, 1.01; $R = 8.314$ J K^{-1} mol^{-1} $= 0.08314$ dm^3 bar K^{-1} mol^{-1})

(a) 2.16 g dm^{-3} (b) 0.072 g dm^{-3} (c) 7.28 g dm^{-3} (d) 216 g dm^{-3}

6. For the reaction, $2SO_{2(g)} + O_{2(g)} \rightarrow 2SO_{3(g)}$, what is the value of $\Delta \nu_g$?

   (a) +5 (b) +1 (c) 0 (d) −1

Chapter 2

Thermodynamics I: Internal Energy Enthalpy, First Law of Thermodynamics, State Functions, and Hess's Law

INTRODUCTION

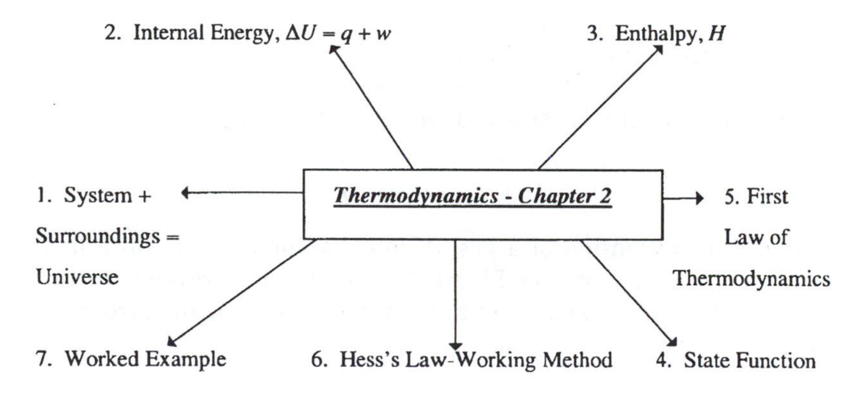

Figure 2.1 *Summary of the contents of Chapter 2*

DEFINITION OF THERMODYNAMICS—INTRODUCING THE SYSTEM AND THE SURROUNDINGS

Chemical thermodynamics is the study of the energy transformations that occur during chemical and physical changes. The part of the universe that undergoes such a change is termed the ***system***. Everything else outside the system is termed the ***surroundings***. Hence, system + surroundings = universe.

INTERNAL ENERGY, U

Energy is the ability or capacity to do work. The unit of energy is the joule, J. The internal energy, U, is defined as the total energy of all forms within a system. Energy is transferred into or out of a system in two ways: (a) by heat transfer, q, and/or (b) by work, w, as shown in Figure 2.2.

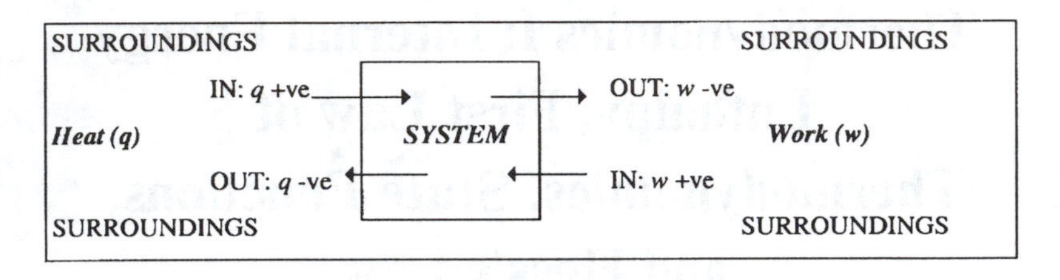

Figure 2.2 *Transfer of energy into and out of a system*

If the work is done by the system ***on*** the surroundings, w is $-$ve, whereas if work is done by the surroundings on the system, w is $+$ve.

Hence, by definition, $$\Delta U = q + w$$

The change in internal energy is defined as $\Delta U = U_{\text{final}} - U_{\text{initial}}$.

ENTHALPY, H

Consider the expansion of a gas, of initial volume, V_{initial}, to a final volume, V_{final}, as shown in Figure 2.3, at constant pressure. In this situation, the gas expands, and hence work is done by the system on its surroundings, *i.e.* w is $-$ve.

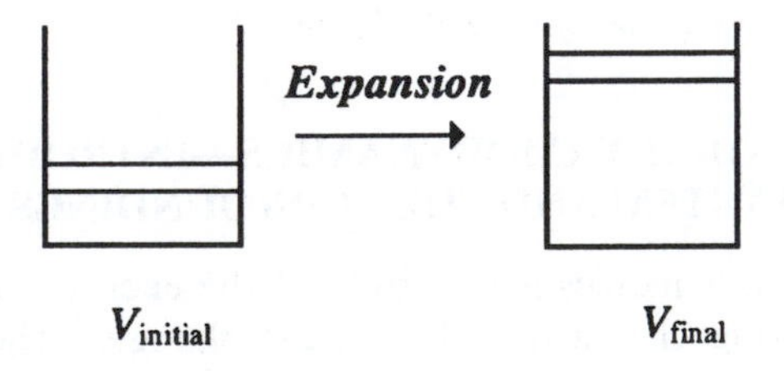

Figure 2.3 *Expansion of a gas. Since the gas expands, the system does work **on** the surroundings, therefore, w is $-$ve*

From physics, work = force × distance ($w = F \times d$) and pressure = force/area ($p = F/A$), as defined in Chapter 1. Therefore, force = pressure × area, *i.e.* $w = (pA)d$. But since volume = length × breadth × height = area × height, this means $w = p\Delta V$, where ΔV is the change in volume, *i.e.* $V_{\text{final}} - V_{\text{initial}}$. Since w is $-$ve, then $w = -p\Delta V$.

But, $\Delta U = q + w$. Therefore, $q = \Delta U - w = \Delta U - (-p\Delta V) = \Delta U + p\Delta V$. At constant pressure, p, for a system with internal energy, U and volume V, the enthalpy $H(= q_{\text{p}})$ is defined as $H = U + pV$. More specifically, ΔH, the change in enthalpy, is defined as $\Delta H = \Delta U + p\Delta V$. This will be discussed further in Chapter 3.

Change in Enthalpy $\Delta H = \Delta U + p\Delta V$

Under standard state conditions (*i.e.* 1 bar pressure and 25 °C = 298 K), ΔH is written as ΔH°.

Some Examples of Reactions

> ***Example:*** Consider the following four reactions at constant pressure. In each case, answer the following questions: (a) Is work done by the system on the surroundings or by the surroundings on the system? (b) What is the sign of w in each case, *i.e.* is w +ve, is w $-$ve or is $w = 0$?

Reaction 1: $H_2O_{(g)} \rightarrow H_2O_{(s)}$.
At constant pressure, $w = -p\Delta V = -p(V_{\text{final}} - V_{\text{initial}})$. But $V_{\text{initial}} > V_{\text{final}}$, from Chapter 1. Hence, ΔV is $-$ve. Therefore, w is +ve, *i.e.* work is done by the surroundings on the system.

Reaction 2: $2NaNO_{3(s)} + \text{HEAT} \rightarrow 2NaNO_{2(s)} + O_{2(g)}$.
At constant pressure, $w = -p\Delta V = -p(V_{\text{final}} - V_{\text{initial}})$. But $V_{\text{final}} > V_{\text{initial}}$, from Chapter 1. Hence, ΔV is +ve, and therefore, w is $-$ve, *i.e.* work is done by the system on the surroundings.

Reaction 3: $2N_2O_{5(g)} \rightarrow 2N_2O_{4(g)} + O_{2(g)}$.
At constant pressure, $w = -p\Delta v_{\text{g}}RT$, from the equation of state of an ideal gas, Chapter 1. Δv_{g}, is the change in the coefficients of gas reagents = $\sum[v(\textit{Gaseous}\text{ products})] - \sum[v(\textit{Gaseous}\text{ reactants})]$ = (2 + 1) − (2) = (3) − (2) = +1. Hence, w is $-$ve, *i.e.* work is done by the system on the surroundings.

Reaction 4: Freezing of water, *i.e.* $H_2O_{(l)} \rightarrow H_2O_{(s)}$.

At constant pressure, $w = -p\Delta V = -p(V_{final} - V_{initial})$. But $V_{initial}$ is approximately equal to V_{final}, from Chapter 1. Hence, $\Delta V \approx 0$. Therefore, $w \approx 0$ *i.e.* no work is done. Note that water does in fact expand slightly on freezing, which is very unusual. This is why water pipes burst in winter.

Heats of Reaction

The ***standard molar enthalpy of formation***, ΔH_f°, is defined as the enthalpy change when 1 mole of a substance is formed from its free elements in their standard states (*i.e.* a solid, liquid or gas at 1 bar pressure and 25 °C = 298 K). For example, for the reaction $C_{(s)} + O_{2(g)} \rightarrow CO_{2(g)}$, ΔH_f°, the standard molar enthalpy of formation of carbon dioxide gas is equal to -394 kJ mol^{-1}, *i.e.* $\Delta H_{rxn}^\circ = \Delta H_f^\circ = -394$ kJ mol^{-1}.

The enthalpy of combustion, ΔH_c is defined as the change in enthalpy when one mole of a substance is burnt in excess oxygen gas at 1 bar pressure and 0 °C = 273 K. For example, for the combustion of propane, $C_3H_{8(l)} + 5O_{2(g)} \rightarrow 3CO_{2(g)} + 4H_2O_{(l)}$, $\Delta H_{rxn} = \Delta H_c = -2220$ kJ mol^{-1} (rxn = reaction).

THE FIRST LAW OF THERMODYNAMICS

The First Law of Thermodynamics is the ***law of conservation of energy***, *i.e.* energy cannot be created or destroyed, but is converted from one form to another. Expressed in an alternative way, the First Law of Thermodynamics states that the total energy of the universe is constant, *i.e.* $\Delta U_{universe} = 0$.

STATE FUNCTIONS

When a certain property of a system (such as U, the internal energy) is independent of how that system attains the state that exhibits such a property, the property is deemed to be a ***state function***, *i.e.* it does not matter which path is followed when a system changes from its initial state to its final state; all that is relevant is the value of such a function in its final state. This concept forms the basis of Hess's Law.

HESS'S LAW

Definition of Hess's Law: Hess's Law states that a reaction enthalpy is the sum of the enthalpies of any path into which the reaction may be divided at the same temperature and pressure.

i.e. the change in enthalpy is *independent* of the path followed. For this reason, enthalpy, like internal energy, is also a state function, a quantity whose value is determined only by the state of the system in question.

Figure 2.4 shows an example of the application of Hess's Law. Consider the combustion of carbon (graphite) in oxygen gas to form carbon dioxide. $CO_{2(g)}$ can be formed in two ways: (a) direct combination of elemental carbon with oxygen to form carbon dioxide, or (b) in two stages, first the combustion of carbon in oxygen to form carbon monoxide, $CO_{(g)}$, followed by the burning of $CO_{(g)}$ in oxygen to form carbon dioxide. If ΔH is measured for both pathways, it is found that each pathway involves the same quantity of heat at constant pressure. This is always true for any chemical reaction.

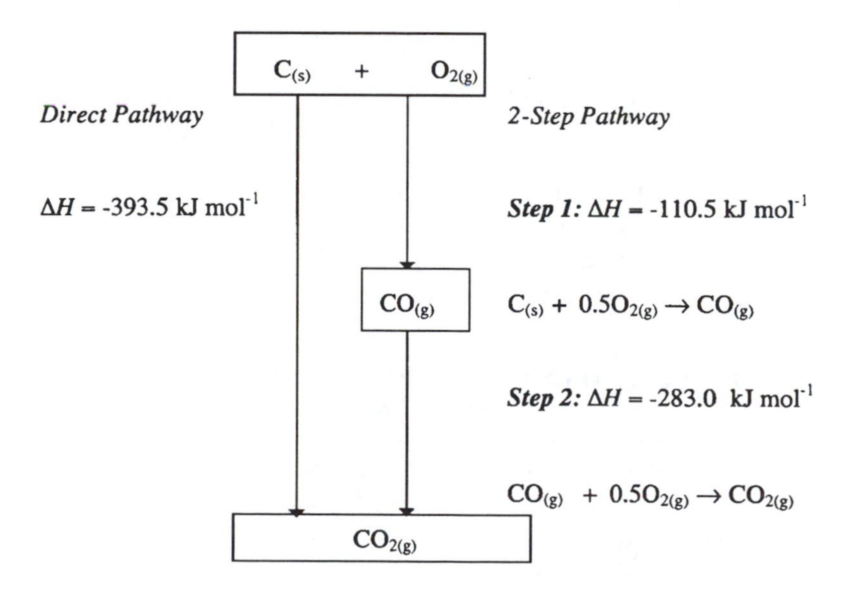

Figure 2.4 *Application of Hess's Law*

In the above example, the sum of the ΔH values of the two-step pathway is equal to the value of ΔH for the direct reaction, *i.e.* $(-110.5) + (-283.0) = -393.5 \text{ kJ mol}^{-1}$.

Working Method on Hess's Law Type Problems

The following working method is a step-by-step procedure on how to determine ΔH_{rxn} using Hess's Law.

1. Read the question carefully.
2. Identify all the species involved, *i.e.* the reactants, the products *and* their states, *i.e.* are they in the solid (s), liquid (l) or gaseous (g) phase? Remember also that *hydrocarbons* (compounds containing hydrogen and carbon only, such as CH_4, C_2H_6 *etc.*) burn in oxygen to form carbon dioxide and water:

   e.g. $CH_{4(g)} + 2O_{2(g)} \rightarrow CO_{2(g)} + 2H_2O_{(l)}$;

   $C_2H_{6(g)} + 3.5O_{2(g)} \rightarrow 2CO_{2(g)} + 3H_2O_{(l)}$.
3. Write a chemical equation for the reaction in question, and balance it.
4. Identify the data given in the question, and write any corresponding formation reactions of the species from their component elements.
5. Identify the unknown in the question, *e.g.* ΔH°_{rxn}, ΔH°_c, *etc.*
6. Examine each of the equations in step 4, and re-arrange them so that the reactants and products required in step 5 are on the same side as those in step 3. However, if the direction of a reaction is changed, the sign of ΔH also changes.
7. Now add the reactions, and find the sum of their enthalpies, by applying Hess's Law, *i.e.* if a reaction is the sum of two or more reactions, then $\Delta H_{rxn} = \Delta H(1) + \Delta H(2) + \ldots$ *etc.*
8. Answer any riders to the question.

WORKED EXAMPLE USING THE WORKING METHOD OF HESS'S LAW

Example: Use Hess's Law to determine the standard enthalpy of combustion of methane (natural gas), $CH_{4(g)}$, given the following data: $\Delta H^\circ_f(CO_{2(g)}) = -393.51$ kJ mol^{-1}, $\Delta H^\circ_f(H_2O_{(l)}) = -285.83$ kJ mol^{-1} and $\Delta H^\circ_f(CH_{4(g)}) = -74.81$ kJ mol^{-1}.

1. Read the question carefully—combustion is involved, and all the data given refer to standard enthalpies of formation!
2. Identify the species involved (*remember hydrocarbons burn in oxygen to form carbon dioxide and water—don't forget this!*):

   $CH_{4(g)}$, $O_{2(g)}$, $CO_{2(g)}$ and $H_2O_{(l)}$.
3. Write a balanced chemical equation for the combustion reaction:

   $CH_{4(g)} + O_{2(g)} \rightarrow CO_{2(g)} + H_2O_{(l)}$ . . . not yet balanced!

   $CH_{4(g)} + 2O_{2(g)} \rightarrow CO_{2(g)} + 2H_2O_{(l)}$ . . . balanced!

4. Identify the data given in the question, and write any corresponding formation reactions of the species from their component elements:

   (a) $C_{(s)} + O_{2(g)} \rightarrow CO_{2(g)}$ $\Delta H^\circ_f(CO_{2(g)}) = -393.51$ kJ

   (b) $2H_{2(g)} + O_{2(g)} \rightarrow 2H_2O_{(l)}$ $\Delta H^\circ_f(H_2O_{(l)}) = (2 \times -285.83) = -571.66$ kJ mol^{-1}

   (c) $C_{(s)} + 2H_{2(g)} \rightarrow CH_{4(g)}$ $\Delta H^\circ_f(CH_{4(g)}) = -74.81$ kJ

5. Identify the unknown in the question.

   $CH_{4(g)} + 2O_{2(g)} \rightarrow CO_{2(g)} + 2H_2O_{(l)}$ $\Delta H^\circ_{rxn} = \Delta H^\circ_c$

6. Examine each of the three equations in step 4, and re-arrange them, such that the reactants and products required in step 5 are on the same side as those in step 3. However, if the direction of a reaction is changed, the sign of ΔH° also changes.

   (a) $C_{(s)} + O_{2(g)} \rightarrow CO_{2(g)}$ $\Delta H^\circ_f(CO_{2(g)}) = -393.51$ kJ

   No problem here, as $CO_{2(g)}$ is on the right-hand side.

   (b) $2H_{2(g)} + O_{2(g)} \rightarrow 2H_2O_{(l)}$ $\Delta H^\circ_f(H_2O_{(l)}) = -571.66$ kJ

   Neither is there a problem here, as $H_2O_{(l)}$ is also on the right-hand side, and the stoichiometry is correct (*i.e.* 2), so ΔH° does not have to be changed.

   (c) $C_{(s)} + 2H_{2(g)} \rightarrow CH_{4(g)}$ $\Delta H^\circ_f(CH_{4(g)}) = -74.81$ kJ

   In this equation, $CH_{4(g)}$ should be on the left-hand side. Hence, if the direction of the reaction is changed, the sign of ΔH°_f also changes.

   Hence: (c) $CH_{4(g)} \rightarrow C_{(s)} + 2H_{2(g)}$ $\Delta H^\circ_f(CH_{4(g)}) = +74.81$ kJ

7. Now add the three reactions, and find the sum of their enthalpies:

   (a) $\not{C}_{(s)} + O_{2(g)} \rightarrow CO_{2(g)}$ $\Delta H^\circ_f(CO_{2(g)}) = -393.51$ kJ

   (b) $2\not{H}_{2(g)} + O_{2(g)} \rightarrow 2H_2O_{(l)}$ $\Delta H^\circ_f(H_2O_{(g)}) = -571.66$ kJ

   (c) $CH_{4(g)} \rightarrow \not{C}_{(s)} + 2\not{H}_{2(g)}$ $\Delta H^\circ_f(CH_{4(g)}) = +74.81$ kJ

   (d) $CH_{4(g)} + 2O_{2(g)} \rightarrow CO_{2(g)} + 2H_2O_{(l)}$

   $\Delta H^\circ_{rxn} = \Delta H^\circ_c = (-393.51) + (-571.83) + (+74.81) = -890.36$ kJ mol^{-1}.

   The value of ΔH°_c means that when methane is burned in a stream of oxygen gas, 890.36 kJ mol^{-1} of heat is given out. This is an example of an exothermic reaction, since ΔH°_{rxn} is negative. This will be discussed further in Chapter 3.

Chapter 3

Thermodynamics II: Enthalpy, Heat Capacity, Entropy, the Second and Third Laws of Thermodynamics, and Gibbs Free Energy

INTRODUCTION

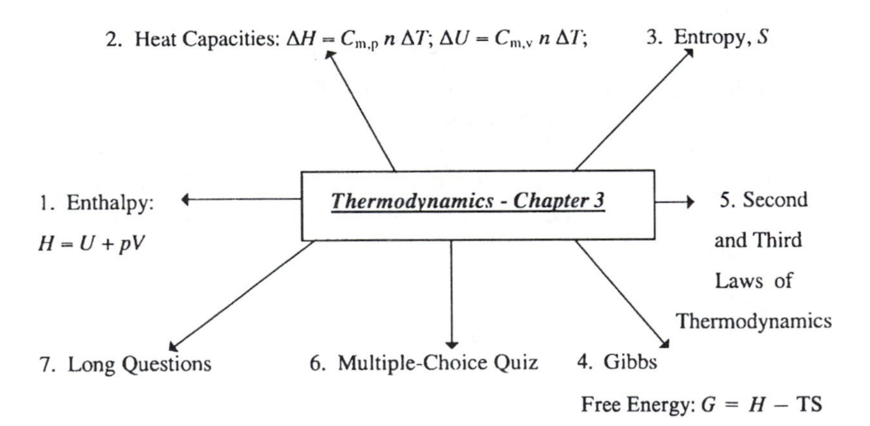

Figure 3.1 *Summary of the contents of Chapter 3*

ENTHALPY, *H*

At constant pressure p, for a system with internal energy U and volume V, the enthalpy is defined as, $H = U + pV$. The change in enthalpy, ΔH, is then expressed as $\Delta H = \Delta U + p\Delta V$. From Chapter

2, we know that the internal energy of a system can be expressed as: $\Delta U = q + w \Rightarrow q = \Delta U - w \Rightarrow q = \Delta U + p\Delta V$. This means that q (or more precisely, q_p, at constant pressure) is the change in enthalpy, ΔH.

Change in Enthalpy, ΔH

The amount of heat, q_p, exchanged when the work done by the system is expansion work at constant pressure is termed ***ΔH, the change in enthalpy*** $(\Delta H = q_p) \Rightarrow \Delta H = H_{final} - H_{initial} = (U_{final} + pV_{final}) - (U_{initial} + pV_{initial})$. Since the pressure is a constant, $\Delta H = (U_{final} - U_{initial}) + p(V_{final} - V_{initial}) = \Delta U + p\Delta V$, where $p\Delta V$ is related to $\Delta\nu_g RT$, from the equation of state of an ideal gas, Chapter 1 $(p\Delta V = \Delta\nu_g RT)$, where $\Delta\nu_g$ = change in the coefficients of gaseous reagents,

$$i.e.\ \Delta\nu_g = \sum[\nu(\textit{Gaseous}\text{ products})] - \sum[\nu(\textit{Gaseous}\text{ reactants})]$$

For example, in the reaction, $C_2H_{6(g)} + 3.5O_{2(g)} \rightarrow 3H_2O_{(l)} + 2CO_{2(g)}$, $\Delta\nu_g = (2) - (3.5 + 1) = -2.5$. If the system is at both constant pressure *and* constant volume $(\Delta V = 0)$, $\Delta H = \Delta U + p\Delta V = \Delta U$.

Summary: $\Delta H = \Delta U + p\Delta V = \Delta U + \Delta\nu_g RT$ at constant pressure; $\Delta H = \Delta U$ at constant volume, since $\Delta V = 0$.

Under standard state conditions, *i.e.* 1 bar pressure and 25 °C (298 K), $\Delta H = \Delta H^\circ$. ΔH°_{rxn} can be easily determined, using the expression:

$$\Delta H^\circ_{rxn} = \sum[\Delta H^\circ_f\ (\text{Products})] - \sum[\Delta H^\circ_f\ (\text{Reactants})]$$

i.e. for the reaction, $\nu_A A + \nu_B B \rightarrow \nu_C C + \nu_D D$, where ν_A, ν_B, ν_C and ν_D are the stoichiometry factors, $\Delta H^\circ_{rxn} = [(\nu_C \Delta H^\circ_{f(C)}) + (\nu_D \Delta H^\circ_{f(D)})] - [(\nu_A \Delta H^\circ_{f(A)}) + (\nu_B \Delta H^\circ_{f(B)})]$.

Changes in the Enthalpy of an Element

The standard molar enthalpy of formation of an element in its most stable state (solid, liquid or gas) is zero, since the formation of an element from itself is not a reaction. Hence, ΔH_f° (element) = 0, *e.g.* $\Delta H^\circ(O_{2(g)}) = 0$, $\Delta H^\circ(Ag_{(s)}) = 0$, *etc.*

$$\boxed{\Delta H_f^\circ(\text{element}) = 0}$$

Exothermic and Endothermic Reactions

The sign of ΔH indicates whether a reaction is exothermic (heat given out, ΔH –ve) or endothermic (heat taken in, ΔH +ve). For example, in the reaction $C_{(s)} + O_{2(g)} \rightarrow CO_{2(g)}$, ΔH_{rxn}° is equal to -393.5 kJ mol^{-1}. Since ΔH_{rxn}° is negative, this means that heat is given out in this reaction, and hence the burning of coal is an exothermic reaction.

ΔH –ve ***exothermic reaction***; ΔH +ve ***endothermic reaction***

Example

Example: Determine the standard heat of combustion of the sugar, sucrose ($C_{12}H_{22}O_{11(s)}$), to form $CO_{2(g)}$ and $H_2O_{(l)}$, given that $\Delta H_f^\circ(C_{12}H_{22}O_{11(s)}) = -2219$ kJ mol^{-1}, $\Delta H_f^\circ(CO_{2(g)}) = -393.5$ kJ mol^{-1} and $\Delta H_f^\circ(H_2O_{(l)}) = -285.8$ kJ mol^{-1}.

Solution:

1. Identify the species present—$C_{12}H_{22}O_{11(s)}$, $CO_{2(g)}$, $H_2O_{(l)}$ and $O_{2(g)}$, since combustion means burning in oxygen gas!
2. Write a balanced chemical equation for the reaction:

   $C_{12}H_{22}O_{11(s)} + O_{2(g)} \rightarrow CO_{2(g)} + H_2O_{(l)}$ . . . not yet balanced!

   $C_{12}H_{22}O_{11(s)} + 12O_{2(g)} \rightarrow 12CO_{2(g)} + 11H_2O_{(l)}$ . . . now balanced!
3. Ensure that all units are the same (all kJ mol^{-1}) so can proceed to step 4.
4. Apply the formula:

   $\Delta H_c^\circ = \sum[\Delta H_f^\circ(\text{Products})] - \sum[\Delta H_f^\circ(\text{Reactants})]$

   $= [12\Delta H_f^\circ(CO_{2(g)}) + 11\Delta H_f^\circ(H_2O_{(l)})] - [1\Delta H_f^\circ(C_{12}H_{22}O_{11(s)})$

   $+12\Delta H_f^\circ(O_{2(g)})]$.
5. Check to see whether any of the species are elements. If so, $\Delta H_f^\circ = 0$. Oxygen is an element, hence $\Delta H_f^\circ(O_{2(g)}) = 0$.

6. Substitute the values, including the stoichiometry factors:
   $\Delta H^\circ_c = \{[12 \times (-393.5) + 11 \times (-285.8)]\} - \{[1 \times (-2219) + 12 \times (0)]\} = -5646.8 \text{ kJ mol}^{-1}$.

Answer: $\Delta H^\circ_c = -5646.8 \text{ kJ mol}^{-1}$

HEAT CAPACITY

The ***heat capacity*** of a body, **C**, is the amount of heat required to raise the temperature of that body by 1 kelvin. The ***specific heat capacity***, **c**, is the amount of heat required to raise the temperature of 1 gram of a body by 1 kelvin, *i.e.* $C = c \times$ molar mass (M). In general,
heat lost or gained = mass × specific heat capacity × change in temperature of a body, *i.e.* $q = mc\Delta T$, from which $c = q/(m\Delta T)$.

Summary:	Heat capacity	$C = q/\Delta T$	J K^{-1}
	Specific heat capacity	$c = q/(m\Delta T)$	$\text{J g}^{-1}\text{ K}^{-1}$

Therefore, $C_{m,p}$ is the amount of heat required to raise the temperature of *1 mole* of a substance by 1 K, at *constant pressure*, and $C_{m,v}$ is the amount of heat required to raise the temperature of *1 mole* of a substance by 1 K, at *constant volume*. From this, two important equations can be derived:

1. At constant pressure: $\Delta U = q_p + w_p$.
But, the work done by the system on the surroundings (expansion work) at constant pressure is $-p\Delta V$, as shown in Chapter 2.
$\Rightarrow \Delta U = q_p - p\Delta V \Rightarrow q_p = \Delta U + p\Delta V = \Delta H$. But, since $q_p = nC_{m,p}\Delta T$, this must equal ΔH, *i.e.* $q_p = \Delta H = nC_{m,p}\Delta T$.

2. At constant volume and constant pressure: $\Delta H = \Delta U + p\Delta V = \Delta U$, since $\Delta V = 0$.
Hence, $q_v = nC_{m,v}\Delta T = \Delta U$.

Summary: 1. At constant pressure: $q_p = \Delta H = nC_{m,p}\Delta T$
2. At constant volume: $q_v = \Delta U = nC_{m,v}\Delta T$

Example No. 1: When a flask containing 500 g of water is heated, the temperature of the water increases from 25 °C to 75 °C. Determine the amount of heat the water absorbs, given that the specific heat capacity of water is $4.184 \text{ J g}^{-1}\text{K}^{-1}$.

Solution:
$q = mc\Delta T = mc(T_{final} - T_{initial}) = (500 \times 4.184) \times (348-298) = 104\,600$ J $= 104.6$ kJ. Notice that the sign of q is +ve. This means that the water has absorbed heat.

> ***Example No. 2:*** Show that for an ideal gas, $C_{m,p} = C_{m,v} + R$

Proof: At constant pressure: $\Delta H = \Delta U + p\Delta V$
$\Rightarrow nC_{m,p}\Delta T = nC_{m,v}\Delta T + p\Delta V = nC_{m,v}\Delta T + nR\Delta T$, since from the equation of state of an ideal gas, $p\Delta V = nR\Delta T$.

But, for 1 mole of an ideal gas, $n = 1$.
$\Rightarrow (1)C_{m,p}\Delta T = (1)C_{m,v}\Delta T + (1)R\Delta T$. Then, dividing across by ΔT, $C_{m,p} = C_{m,v} + R$.

In the above expression, $C_{m,p}$ has to be greater than $C_{m,v}$, since a certain amount of heat is needed to increase the temperature at constant volume. However, at constant pressure, more heat is required to raise the temperature and do the work, expanding the volume of the gas (Chapter 1—Charles' Law: $V \propto T$, for a fixed mass of gas at constant pressure).

Example on Heat Capacities

> ***Example:*** $C_{m,p}(Au) = 25.4$ J K^{-1} mol^{-1} and $C_{m,p}(H_2O) = 75.3$ J K^{-1} mol^{-1}. If a gold stud initially at a room temperature of 25 °C is dropped into 12 g of water at 8 °C, the final temperature of the water reaches 13 °C. Given that the molar masses of Au, H and O are 197, 1 and 16 g mol^{-1} respectively, determine the mass of the stud.

Solution:

1. Read the question carefully.
2. Identify the data:
   $M(H_2O) = 2(1) + 16 = 18$ g mol^{-1}; $M(Au) = 197$ g mol^{-1}; 12 g of H_2O
   $C_{m,p}(Au) = 25.4$ J K^{-1} mol^{-1}; $C_{m,p}(H_2O) = 75.3$ J K^{-1} mol^{-1};
   $T_{initial}(Au)$ (K) $= T(°C) + 273 = (25 + 273)$ K $= 298$ K;
   $T_{final}(Au)$ (K) $= (13 + 273)$ K $= 286$ K;
   $T_{initial}(H_2O)$ (K) $= (8 + 273) = 281$ K;
   $T_{final}(H_2O)$ (K) $= (13 + 273) = 286$ K.

3. Draw a simple diagram to visualise the problem:

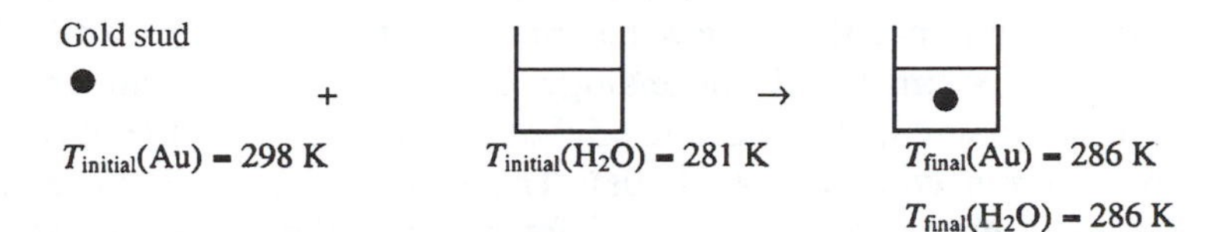

4. Write down the appropriate equations:
   Mass (in g) $m = nM$; $C_{m,p} = q_p/(n\Delta T)$
5. Substitute the values:
   Au: $q_p = n(Au)C_{m,p}(Au)\Delta T$; H_2O: $q_p = n(H_2O)C_{m,p}(H_2O)\Delta T$, where $n = m/M$.
6. Manipulate the equations: heat lost by Au = heat gained by H_2O
   i.e. $n(Au)C_{m,p}(Au)\Delta T = n(H_2O)C_{m,p}(H_2O)\,\Delta T$
   $\Rightarrow -\,[(m/197) \times 25.4 \times (286 - 298)] = +\,[(12/18) \times (75.3) \times (286 - 281)]$
   Note the signs on both sides: heat loss is $-$*ve and heat gain is* $+$*ve.*
   $\Rightarrow m = [(75.3 \times 5 \times 197 \times 12)/(25.4 \times 12 \times 18)] = 162.2$ g.

Answer: Mass = 162.2 g

ENTROPY, *S*, AND CHANGE IN ENTROPY, ΔS

Entropy, *S*, is a measure of the disorder of a system. In Chapter 1, the three states of matter—solid, liquid and gas—were introduced. As you change from the highly regular and ordered solid state to the disordered gaseous state, the disorder or entropy, *S*, increases (Figure 3.2).

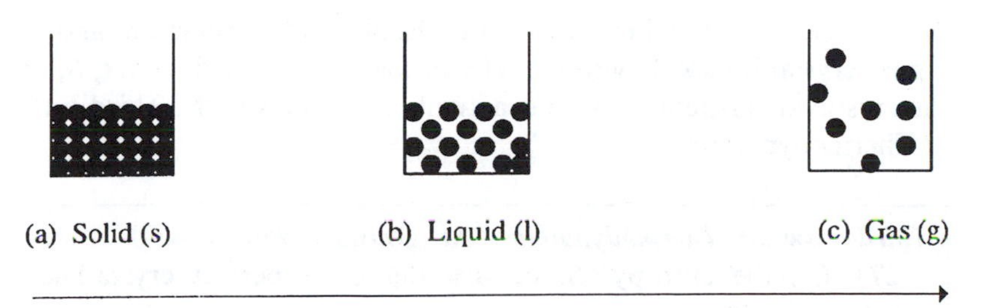

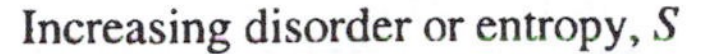

Figure 3.2 *Entropy, S*

Consider the situation in a lecture theatre, when the lecturer leaves for 10 minutes. What is the result? Chaos and disorder! The class will not maintain the same order as that assumed when the lecturer is present! Such an analogy is basically the Second Law of Thermodynamics, *i.e.* the entropy or disorder, S, of the universe tends to a maximum (*i.e.* chaos). The first two laws of thermodynamics (Law of Conservation of Energy and the tendency for the entropy of the universe to increase to a maximum) can be summarised as follows:

First and Second Laws of Thermodynamics: The energy of the universe is constant (First Law) and the entropy, S, of the universe tends to a maximum (Second Law).

Consider the following three examples:

(a) $N_2O_{3(g)} \rightarrow NO_{(g)} + NO_{2(g)}$. Here, 1 mole of gas → 2 moles of gas. Hence, the disorder has increased, *i.e.* ΔS is +ve.

(b) $NH_{3(g)} + HCl_{(g)} \rightarrow NH_4Cl_{(s)}$. Here there is a change of state from the highly disordered gaseous state to the more ordered crystalline solid state. Hence, the disorder has decreased, *i.e.* ΔS is −ve.

(c) $H_2O_{(s)} \rightarrow H_2O_{(l)}$. A liquid is more disordered than a solid; hence, ΔS is +ve for this reaction.

ΔS is defined as the change in entropy. ΔS° (at standard state conditions) can be determined in the same way as ΔH° is evaluated:

$$\Delta S^\circ = \sum[S^\circ(\text{Products})] - \sum[S^\circ(\text{Reactants})]$$

The Third Law of Thermodynamics

At 0 K, the vibrational motion of a molecule is at a minimum, and in a pure crystalline solid, with no defects, the entropy or disorder, S, of the crystal at this temperature is actually zero. This is the Third Law of Thermodynamics.

Third Law of Thermodynamics: At absolute zero (*i.e.* 0 K or −273 °C), the entropy, S, or disorder of a perfect crystalline substance is zero.

This concept is shown in Figure 3.3.

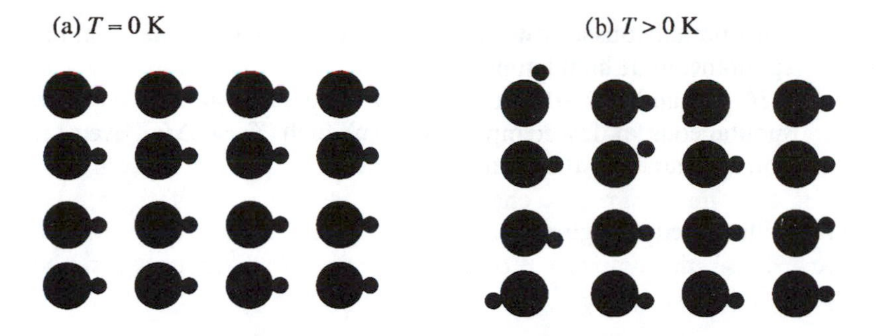

Figure 3.3 *(a) A perfect crystalline solid, AB, at 0* K *(S = 0), and (b) at above 0* K*, when the molecules start to vibrate. The regular array now becomes slightly disordered, i.e. S increases. This is an illustration of the Third Law of Thermodynamics*

GIBBS FREE ENERGY, *G*, AND CHANGE IN GIBBS FREE ENERGY, ΔG

ΔH and ΔS can be combined to give another state function, ΔG, which is the change in Gibbs Free Energy. *G*, the Gibbs Free Energy, is defined as:

> ***Gibbs Free Energy:*** $G = H - TS$ (remembered by ***G***ibbs ***H***a***TS***)

ΔG is a measure of the spontaneity of a reaction, *i.e.*

> ΔG −ve for a spontaneous reaction;
> ΔG +ve for a non-spontaneous reaction;
> $\Delta G = 0$ for a reaction at equilibrium (discussed in Chapter 4)

As ΔH and ΔS can assume both +ve and −ve values, this generates four possible combinations: $\Delta G = \Delta H - T\Delta S$

1. ΔH +ve and ΔS −ve $\Rightarrow$ ΔG +ve, *i.e.* non-spontaneous at all temperatures.
2. ΔH −ve and ΔS +ve $\Rightarrow$ ΔG −ve, *i.e.* spontaneous at all temperatures.
3. The other two combinations, ΔH +ve/ΔS +ve and ΔH −ve/ΔS −ve respectively, are temperature dependent.
   ΔH +ve and ΔS +ve (a) low $T \Rightarrow T\Delta S$ small $\Rightarrow$ ΔG +ve, *i.e.*

non-spontaneous at low temperature; (b) high $T \Rightarrow \Delta G$ −ve, *i.e.* spontaneous at high temperature.

4. ΔH −ve and ΔS −ve (a) low $T \Rightarrow T\Delta S$ small $\Rightarrow \Delta G$ −ve, *i.e.* spontaneous at low temperature; (b) high $T \Rightarrow \Delta G$ +ve, *i.e.* non-spontaneous at high temperature.

This is illustrated in Figure 3.4.

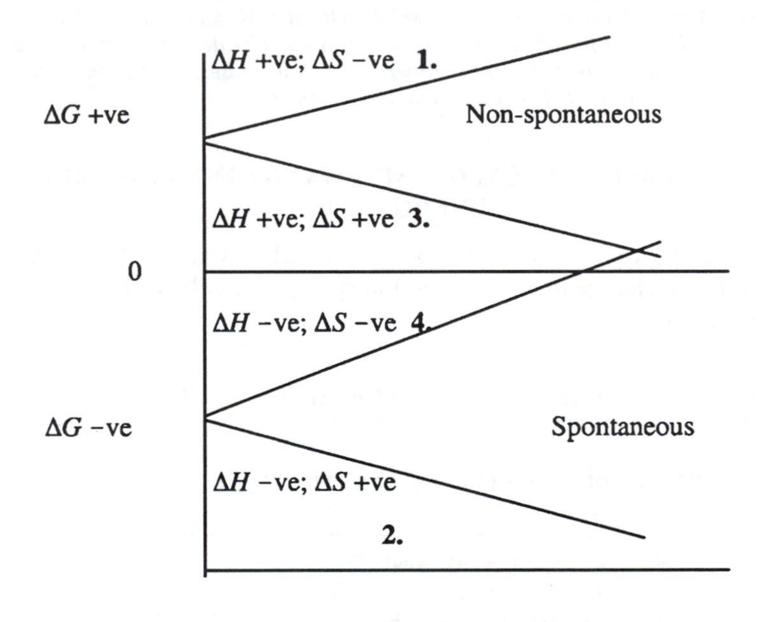

Figure 3.4 *The dependence of* ΔG *and the spontaneity of a reaction on temperature, showing the four possible combinations of* ΔH *and* ΔS *respectively*

ΔG° can be easily calculated, using the expression:

$$\Delta G^\circ_{rxn} = \sum[\Delta G^\circ_f(\text{Products})] - \sum[\Delta G^\circ_f(\text{Reactants})]$$

Note: ΔG°_f (element) = 0.

WORKING METHOD FOR THE CALCULATION OF ΔG TYPE PROBLEMS

A common problem on examination papers is, when given a set of ΔH° and S° data, you are asked to evaluate ΔG°_{rxn}. The following working method describes a skeleton step-by-step outline on how to approach such a problem.

1. Read the question carefully.
2. Identify the species involved (the reactants and the products) and identify their states, *i.e.* (*) = (s), (l) or (g).
3. Write down a balanced chemical equation, with the states indicated, *i.e.* $\nu_A A_{(*)} + \nu_B B_{(*)} \rightarrow \nu_C C_{(*)} + n_D D_{(*)}$, where ν_A, ν_B, ν_C and ν_D, are the stoichiometry factors.
4. Determine $\Delta H^\circ_{rxn} = [(\nu_C \times \Delta H^\circ_{f(C)}) + (\nu_D \times \Delta H^\circ_{f(D)})] - [(\nu_A \times \Delta H^\circ_{f(A)}) + (\nu_B \times \Delta H^\circ_{f(B)})]$, *i.e.* $\Delta H^\circ_{rxn} = \sum[\Delta H^\circ_f(\text{Products})] - \sum[\Delta H^\circ_f(\text{Reactants})]$. Remember $\Delta H^\circ_f(\text{element}) = 0$, and do not forget the units.
5. Determine ΔS°_{rxn} in a similar fashion: $\Delta S^\circ_{rxn} = \sum[S^\circ(\text{Products})] - \sum[S^\circ(\text{Reactants})]$, but S°(element) is ***not*** equal to 0! Write down the units of ΔS°.
6. Determine the temperature, T in K. Remember $T(\text{K}) = [T(^\circ\text{C}) + 273]$ K.
7. Convert ΔH° and ΔS° to the same system of units, *i.e.* ΔH° in J mol^{-1} and ΔS° in J K^{-1} mol^{-1} *or* ΔH° in kJ mol^{-1} and ΔS° in kJ K^{-1} mol^{-1}.
8. Determine the value of ΔG°, using the equation $\Delta G^\circ = \Delta H^\circ - T\Delta S^\circ$, *i.e.* '***G****ibbs* ***H****a****TS***'!
9. Answer any riders to the question. For example: ΔH° −ve, exothermic reaction; ΔH° +ve, endothermic reaction; ΔS° −ve, decrease in the entropy or the disorder of the reaction; ΔS° +ve, increase in the entropy or the disorder of the reaction; ΔG° −ve, spontaneous reaction; ΔG° +ve, non-spontaneous reaction; $\Delta G = 0$ reaction at equilibrium (explained in Chapter 4), where K is defined as the equilibrium constant.
10. At equilibrium,

$$\boxed{\Delta G = \Delta G^\circ + RT\ln K = 0} \qquad \Rightarrow \ln K = -\Delta G^\circ/(RT).$$

WORKED EXAMPLE

Example: Determine $\Delta G°$ for the reaction of $N_{2(g)}$ and $H_{2(g)}$ to form $NH_{3(g)}$, given the following data. Comment on the sign of $\Delta G°$.

	$N_{2(g)}$	$H_{2(g)}$	$NH_{3(g)}$
ΔH_f°/kJ mol^{-1}	0	0	−46.11
$S°$/J K^{-1} mol^{-1}	191.61	130.684	192.45

Solution:

1. Read the question carefully.
2. Species involved: $N_{2(g)}$, $H_{2(g)}$ and $NH_{3(g)}$.
3. Balanced chemical equation: $0.5N_{2(g)} + 1.5H_{2(g)} \rightarrow NH_{3(g)}$.
4. $\Delta H^\circ_{rxn} = [1 \times (-46.11)] - [0.5 \times (0) + 1.5 \times (0)] = -46.11$ kJ mol^{-1}.
5. $\Delta S^\circ_{rxn} = [1 \times (192.45)] - [0.5 \times (191.61) + 1.5 \times (130.684)] = -99.381$ J K^{-1} mol^{-1}.
6. $\Delta G° = \Delta H° - T\Delta S°$
   $= (-46\,110) - [(298) \times (-99.381)]$
   $= -16\,494.462$ J mol^{-1} $= -16.49$ kJ mol^{-1}.
7. $\Delta G°$ −ve ⇒ spontaneous reaction.

SUMMARY OF CHAPTERS 2 AND 3 ON CHEMICAL THERMODYNAMICS

The multiple-choice test and the accompanying three longer questions which follow act as a revision of both Chapters 2 and 3 on thermodynamics.

MULTIPLE-CHOICE TEST

1. A gas expands from a volume of 2.5 dm^3 to 3.7 dm^3 against an external pressure of 1.5 bar, while absorbing 78 J of heat. What is the change in the internal energy of the gas (in joules), given that 1 dm^3 bar = 99.98 J?
   (a) +198.0 (b) +258.0 (c) −101.96 (d) −258.0
2. Which of the following four combinations guarantees that a reaction will proceed spontaneously:

(a) ΔH +ve, ΔS −ve (b) ΔH −ve, ΔS +ve
(c) ΔH +ve, ΔS +ve (d) ΔH −ve, ΔS −ve

3. How many of the following equations are incorrect?
   $\Delta U = nC_{m,p}\Delta T$ $\Delta G = RT\ln K_p$ $\Delta U = \Delta H + p\Delta V$ $\Delta H = nC_{m,v}\Delta T$
   (a) All four (b) three (c) two (d) one
4. Given that for NH_3, ΔH°_{vap} = 23.3 kJ mol^{-1} and ΔS°_{vap} = 92.2 J K^{-1} mol^{-1}, what is the boiling point of NH_3 in K?
   (a) 21 (b) 525 (c) 293 (d) 253
5. Given that $\Delta H^{\circ}_f(FeCl_{2(s)})$ = −341.8 kJ mol^{-1} and $\Delta H^{\circ}_f(FeCl_{3(s)})$ = −399.49 kJ mol^{-1}, what is ΔH°_{rxn} (in kJ mol^{-1}) for the reaction: $FeCl_{2(s)} + 0.5Cl_{2(g)} \rightarrow FeCl_{3(s)}$?
   (a) −57.7 (b) +714.3 (c) +57.7 (d) 0

LONG QUESTIONS ON CHAPTERS 2 AND 3

1. Use Hess's Law to determine ΔH°_{rxn} for the burning of ethanol to form carbon dioxide and water, given the following data: $\Delta H^{\circ}_f(CH_3CH_2OH_{(l)})$ = −277.7 kJ mol^{-1}, $\Delta H^{\circ}_f(CO_{2(g)})$ = −393.51 kJ mol^{-1} and $\Delta H^{\circ}_f(H_2O_{(l)})$ = −285.83 kJ mol^{-1}.
2. Determine ΔG°_{rxn} for the reaction $CH_{4(g)} + N_{2(g)} \rightarrow HCN_{(g)} + NH_{3(g)}$ from the following data:

	$CH_{4(g)}$	$N_{2(g)}$	$HCN_{(g)}$	$NH_{3(g)}$
ΔH°_f/kJ mol^{-1}	−74.81	0	135	−46.11
S°/J K^{-1} mol^{-1}	186.15	191.5	201.7	192.3

   Comment on the significance of the value of ΔG°_{rxn}.
3. Calculate ΔG°_{rxn} for the oxidation of $B_{(s)}$ to $B_2O_{3(s)}$, given the following data:

	$O_{2(g)}$	$B_{(s)}$	$B_2O_{3(s)}$
ΔH°_f/kJ mol^{-1}	0	0	−1272.8
ΔS°/J K^{-1} mol^{-1}	205.03	5.86	53.97

   Comment on the significance of the value of ΔG°_{rxn}.

Chapter 4

Equilibrium I: Introduction to Equilibrium and Le Châtelier's Principle

INTRODUCTION

When reactants combine in a chemical reaction to form products, the conversion of reactants to products is often incomplete, no matter how long the reaction is allowed to continue. In the initial state, the reactants are present at a definite concentration. As the reaction proceeds, the concentration of reactants decreases and after a certain time, t, the concentrations of the reactants level off and become

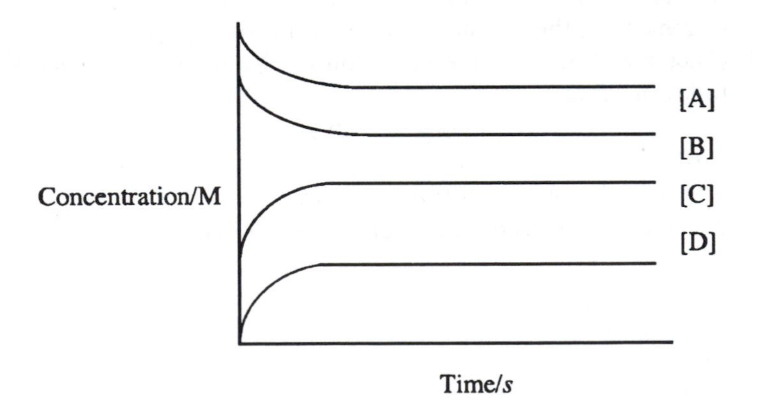

Figure 4.1 *Concentration of reactants and products as a function of time for the reaction* $A + B \rightleftharpoons C + D$

constant (Figure 4.1). A state of ***equilibrium*** is established when the concentrations of reactants (in this example, [A] and [B]) and of products ([C] and [D]) remain constant. At this point, the rate of the forward reaction equals the rate of the reverse reaction.

Once the state of equilibrium is established, it will persist indefinitely, if the system is undisturbed.

> ***Definition of Equilibrium:*** Equilibrium is defined as a state of dynamic balance when two opposing reactions occur at the same time and the same rate.
>
> *i.e.* Rate of forward reaction = rate of backward reaction.

THE LAW OF CHEMICAL EQUILIBRIUM

Every chemical reaction has its own state of equilibrium, in which there is a definite relationship between the concentrations of reactants and products in the reaction. If the conditions of the reaction are varied (*e.g.* if different initial concentrations are used), the concentrations at equilibrium will be changed, but will still assume a constant value. If the concentrations at equilibrium are expressed in mol dm^{-3} (M), there is a single expression which holds for all experimental conditions for a given reaction.

For example, for the reaction, $A + B \rightleftharpoons C + D$, at equilibrium:

$K_c = \dfrac{[C][D]}{[A][B]}$, where K_c is termed ***the equilibrium constant*** and is characteristic of a given reaction.

In general, for the reaction: $\nu_A A + \nu_B B \rightleftharpoons \nu_C C + \nu_D D$, where ν_A, ν_B, ν_C and ν_D now represent the stoichiometry factors,

$$K_c = \frac{[C]^{\nu_C}[D]^{\nu_D}}{[A]^{\nu_A}[B]^{\nu_B}}$$

i.e. of the form 'products/reactants'.
For the reaction

$$I_{2(g)} + H_{2(g)} \rightleftharpoons 2HI_{(g)},$$

$$K_c = \frac{[HI]^2}{[I_2]^1[H_2]^1}$$

Partial Pressures

When gases are involved in a reaction, the partial pressure of each gas has to be considered. For the general reaction

$$A_{(g)} + B_{(g)} \rightleftharpoons C_{(g)} + D_{(g)},$$

$$K_p = \frac{\{p(C)/p^{\ddagger}\}\{p(D)/p^{\ddagger}\}}{\{p(A)/p^{\ddagger}\}\{p(B)/p^{\ddagger}\}}$$

where p is the partial pressure of the gas and $p^{\ddagger}$ is the standard pressure. K_p denotes the equilibrium constant obtained by using the partial pressures of the gases, instead of their concentrations. For this reason, K_p is always dimensionless, as all pressures are divided by the standard pressure. This is implicit in the definition of the standard Gibbs free energy change, $\Delta G°$. Therefore, the above equation can be expressed in a more convenient form as

$$K_p = \frac{\{p(C)'\}\{p(D)'\}}{\{p(A)'\}\{p(B)'\}}$$

where $p(A)' = p(A)/p^{\ddagger}, p(B)' = p(B)/p^{\ddagger}$, *etc.*

For simplicity, the prime will be dropped in subsequent equations involving partial pressures, but it is always implied. For example, for the reaction involving the manufacture of ammonia by the Haber process,

$$N_{2(g)} + 3H_{2(g)} \rightleftharpoons 2NH_{3(g)},$$

$$K_p = \frac{p(NH_3)^2}{\{p(N_2)^1\}\{p(H_2)^3\}}$$

But, K_p is related to K_c: for the reaction

$$\nu_A A_{(g)} + \nu_B B_{(g)} \rightleftharpoons \nu_C C_{(g)} + \nu_D D_{(g)}$$

$$K_p = \frac{\{p(C)\}^{\nu_C}\{p(D)\}^{\nu_D}}{\{p(A)\}^{\nu_A}\{p(B)\}^{\nu_B}}$$

But, from Chapter 1, the equation of state of an ideal gas is defined as:

$pV = nRT \Rightarrow p = (nRT)/V$, where $n =$ amount of gas (expressed in moles)

$$\Rightarrow \quad K_p = \frac{\{(n_C RT)/V_C\}^{\nu_C}\{(n_D RT)/V_D\}^{\nu_D}}{\{n_A RT)/V_A\}^{\nu_A}\{n_B RT/V_B\}^{\nu_B}}$$

$$K_p = \frac{\{(n_C/V_C\}^{\nu_C}(RT)^{\nu_C}\{n_D/V_D\}^{\nu_D}(RT)^{\nu_D}}{\{(n_A/V_A\}^{\nu_A}(RT)^{\nu_A}\{(n_B/V_B\}^{\nu_B}(RT)^{\nu_B}}$$

$$\Rightarrow \quad K_p = \{[C]^{\nu_C}[D]^{\nu_D}/[A]^{\nu_A}[B]^{\nu_B}\}\{(RT)\}^{(\nu_C+\nu_D)-(\nu_A+\nu_B)}$$

since concentration $[X = (n_X/V_X)$ is expressed as the number of moles per cubic decimetre.

$$\Rightarrow \quad K_p = K_c(RT)^{\Delta\nu_g}$$

where $\Delta\nu_g =$ change in the coefficients of ***gaseous*** reagents
i.e. ν_g (*gaseous* products) $-$ ν_g (*gaseous* reactants).

This is a very important relationship, which relates K_p to K_c. *Do not, however, mix up the order;* this can be remembered by *PC—'political correctness' (not the other way around!).*
Note: While the units of K_c are always derived from the equations as stated previously, *e.g.*:

(a) $H_{2(g)} + I_{2(g)} \rightleftharpoons 2HI_{(g)}, K_c = \dfrac{[HI]^2}{[I_2]^1[H_2]^1}$

i.e. $M^2/(M\,M) \Rightarrow$ no units;

(b) $N_{2(g)} + 3H_{2(g)} \rightleftharpoons 2NH_{3(g)}, K_c = \dfrac{[NH_3]^2}{[N_2]^1[H_2]^3}$

i.e. $M^2/(M\,M^3) \Rightarrow$ units of M^{-2};

K_p is always dimensionless, as all pressures are divided by the standard pressure, *e.g.* in the latter reaction:

$$K_p = \frac{\{p(NH_3)/p^{\ominus}\}^2}{\{p(N_2)/p^{\ominus}\}^1\{p(H_2)/p^{\ominus}\}^3}$$

i.e. $(\text{bar/bar})^2/(\text{bar/bar})^1\,(\text{bar/bar})^3 \Rightarrow$ dimensionless!

Since the partial pressures of gases are normally expressed in bars, $R = 0.08314\ \text{dm}^3\ \text{bar}\ \text{K}^{-1}\ \text{mol}^{-1}$ should always be used in K_p

problems, and not $R = 8.314\ \mathrm{J\ K^{-1}\ mol^{-1}}$,
e.g. for the reaction

$$2NOCl_{(g)} \rightleftharpoons Cl_{2(g)} + 2NO_{(g)}\quad K_c = 3.75 \times 10^{-6} \text{ at } 1069\ \mathrm{K}$$

$\Rightarrow K_p = K_c\,(RT)^{\Delta\nu_g} = (3.75 \times 10^{-6}) \times \{(0.08314\ \mathrm{dm^3\ bar\ K^{-1}\ mol^{-1}}) \times (1069\ \mathrm{K})\}^1$,
$\Delta\nu = (1 + 2) - (2) = 1$, where $K_p = \{p(Cl_2)^1\ p(NO)^2\}/p(NOCl)^2$.
Hence, $K_p = 3.33 \times 10^{-4}$.

Homogeneous and Heterogeneous Equilibria

(a) Homogeneous equilibrium.—This is an equilibrium involving all species in the *one phase*, *e.g.* all gases, such as $N_{2(g)} + 3H_{2(g)} \rightleftharpoons 2NH_{3(g)}$.
(b) Heterogeneous equilibrium.—This is an equilibrium when *more than one phase* is involved in the process, *e.g.* if calcium carbonate (limestone or marble) is heated, carbon dioxide gas is evolved [which can be detected by bubbling it through calcium hydroxide, $Ca(OH)_2$ (limewater), which turns milky], *i.e.* $CaCO_{3(s)} + \text{HEAT} \rightleftharpoons CaO_{(s)} + CO_{2(g)}$. Two phases are involved here, the solid phase and the gaseous phase, and hence this is an example of a heterogeneous equilibrium reaction, where $K_c = \{[CaO_{(s)}][CO_{2(g)}]\}/[CaCO_{3(s)}]$. But since the concentration of a solid is constant, this means $K_c = [CO_{2(g)}]$ and $K_p = p(CO_2)$. In fact, concentrations and pressures are approximations of the activity, a, of a substance, and the activity of a pure solid or pure liquid is unity, *e.g.* $a_{(s)} = 1$.

For the equilibrium $H_2O_{(l)} \rightleftharpoons H_2O_{(g)}$, $K_c = [H_2O_{(g)}]$, since the activity a of pure $H_2O_{(l)} = 1$ and $K_p = p(H_2O)$.

WORKING METHOD FOR THE SOLUTION OF STANDARD EQUILIBRIUM TYPE PROBLEMS

1. Read the question carefully.
2. Identify all species involved, including their states, *i.e.* are they (s), (l) or (g)? Identify the type of equilibrium, *i.e.* homogeneous or heterogeneous equilibrium. Remember the activity, a, of a solid or pure liquid is unity.
3. Write down the balanced chemical equation, with all states included. This may not necessarily be given in the question. This is the most important step in any equilibrium problem, as it dictates the expression for K_c and subsequent equilibrium

concentrations. Hence, it is essential that the equation is correct, with the appropriate stoichiometry factors, ν_A, ν_B, ν_C, ν_D, *etc.*

4. Having checked that the equation is completely balanced in step 3, write down an expression for the equilibrium constant for the reaction of the form,

$$K_c = \text{products/reactants.}$$

i.e. for the reaction $\nu_A A + \nu_B B \rightleftharpoons \nu_C C + \nu_D D$,
$K_c = [C]^{\nu_C}[D]^{\nu_D}/[A]^{\nu_A}[B]^{\nu_B}$.

5. If gases are involved, write down an expression for K_p:

$$K_p = \frac{\{p(C)\}^{\nu_C}\{p(D)\}^{\nu_D}}{\{p(A)\}^{\nu_A}\{p(B)\}^{\nu_B}}$$

and relate K_p to K_c, by determining the change in the coefficents of *gaseous* reagants!
i.e. $\Delta\nu = \nu$ (*gaseous* products) $- \nu$ (*gaseous* reactants)
$K_p = K_c\,(RT)^{\Delta\nu}$
Remember the order: PC—'political correctness'!

6. Convert any concentrations to mol dm^{-3} (M), *etc.* Create a table of the initial and final concentrations at equilibrium, letting x = the change in concentration. This is related to the stoichiometry factors, ν_A, ν_B, ν_C, *etc.*
For example, in the reaction:

	$N_{2(g)}$	+	$3H_{2(g)}$	$\rightleftharpoons$	$2NH_{3(g)}$
***Initial conc.*/M**	0.12		0.25		0
Change	$-x$		$-3x$		$+2x$
***Final conc.*/M**	$0.12 - x$		$0.25 - 3x$		$0 + 2x$

7. Substitute the final equilibrium concentrations into the expression for K_c (step 4), and solve for x.
8. If the change (*i.e.* x) is assumed to be considerably less than the initial concentration, this can be neglected. If x is then found to be $<5\%$ of the initial concentration, the ***assumption made is valid***. If not, a quadratic equation of the form $ax^2 + bx + c = 0$ may need to be solved, with solution, $x = \dfrac{-b \pm \sqrt{b^2 - 4ac}}{2a}$
9. Determine the final equilibrium concentrations.
10. State the units of K_c. Remember also that K_p is dimensionless.

EXAMPLES

> ***Example No. 1:*** Determine the concentration of all species at equilibrium at 973 K, when 2 moles of $O_{2(g)}$ in 4.0 dm^3 are mixed with 2.25 moles of $N_{2(g)}$ in 3.0 dm^3, according to the equilibrium reaction: $O_{2(g)} + N_{2(g)} \rightleftharpoons 2NO_{(g)}$, given that $K_c = 4.13 \times 10^{-9}$. Is this an example of a homogeneous or a heterogeneous equilibrium? Evaluate K_p, given that $R = 0.08314$ dm^3 bar K^{-1} mol^{-1}.

1. Read the question carefully.
2. Identify the species involved: $O_{2(g)}$, $N_{2(g)}$ and $NO_{(g)}$. All species are in the gaseous phase; therefore this is a homogeneous equilibrium (one single phase).
3. The balanced chemical equation is already given in the question:

$$O_{2(g)} + N_{2(g)} \rightleftharpoons 2NO_{(g)}.$$

4. $K_c = [NO]^2/\{[N_2][O_2]\} = 4.13 \times 10^{-9}$.
5. $K_p = p(NO)^2/\{p(N_2)p(O_2)\} = K_c\,(RT)^{\Delta\nu}$, where $\Delta\nu = (2)-(2) = 0$. Hence $K_p = K_c(RT)^0 = 4.13 \times 10^{-9}$, since from the rules of indices, $x^0 = 1$.
6. Initial concentration of $O_{2(g)}$ = 2 mol in 4.0 dm^3 $\Rightarrow$ 0.5 mol in 1 dm^3, *i.e.* 0.5 M
   Initial concentration of $N_{2(g)}$ = 2.25 mol in 3.0 dm^3 $\Rightarrow$ 0.75 mol in 1 dm^3, *i.e.* 0.75 M

	$O_{2(g)}$	+	$N_{2(g)}$	$\rightleftharpoons$	$2NO_{(g)}$
***Initial conc.*/M**	0.5		0.75		0
Change	$-x$		$-x$		$+2x$
***Final conc.*/M**	$0.5 - x$		$0.75 - x$		$0 + 2x$

7. $K_c = [NO]^2/\{[N_2][O_2]\} = 4.13 \times 10^{-9} = (2x)^2/\{(0.5 - x)(0.75 - x)\}$.
8. Since $K_c \ll 1$ (*i.e.* the equilibrium lies to the left), assume $x \ll 0.5$ and $x \ll 0.75 \Rightarrow K_c = (4x^2)/\{(0.5) \times (0.75)\} = 4.13 \times 10^{-9}$. Hence, $x^2 = 3.8719 \times 10^{-10}$ and $x = 1.97 \times 10^{-5}$.
9. $(x/0.5)\% = (1.97 \times 10^{-5}/0.5)\% = 0.00394\%$ and $(x/0.75)\% = (1.97 \times 10^{-5}/0.75)\% = 0.00263\%$ *i.e.*, both are $<5\%$. Hence the assumptions made were valid and it is not necessary to solve a quadratic equation in this question.
10. $[O_{2(g)}] \approx 0.5$ M; $[N_{2(g)}] \approx 0.75$ M (both slightly reduced); $[NO_{(g)}] = 2x = 2 \times (1.97 \times 10^{-5}) = 3.94 \times 10^{-5}$ M.
11. $K_p = 4.13 \times 10^{-9}$.

Example No. 2: When 1 M $H_{2(g)}$ and 1 M $I_{2(g)}$ come to equilibrium at 730 K, determine the concentration of each substance at equilibrium, if $K_c = 48.9$. Determine the value of K_p given that $R = 0.08314$ dm^3 bar K^{-1} mol^{-1}.

1. Read the question carefully.
2. Species involved: $H_{2(g)}$, $I_{2(g)}$ and $HI_{(g)}$, *i.e.* all gases ⇒ homogeneous equilibrium (one single phase).
3. No balanced chemical equation is given. Therefore:
   (a) identify the reactants: $H_{2(g)}$ and $I_{2(g)}$;
   (b) identify the product: $HI_{(g)}$;
   (c) write a chemical equation for the equilibrium reaction and balance:

$$H_{2(g)} + I_{2(g)} \rightleftharpoons 2HI_{(g)}.$$

4. $K_c = [HI]^2/\{[H_2][I_2]\} = 48.9$.
5. $K_p = p(HI)^2/\{p(H_2)p(I_2)\} = K_c\,(RT)^{\Delta\nu_g}$, where $\Delta\nu_g = (2) - (2) = 0$.
   Hence $K_p = K_c(RT)^0 = 48.9$.
6.

	$H_{2(g)}$	+	$I_{2(g)}$	$\rightleftharpoons$	$2HI_{(g)}$.
Initial conc./M	1		1		0
Change	$-x$		$-x$		$+2x$
Final conc./M	$1 - x$		$1 - x$		$0 + 2x$

7. $K_c = [HI]^2/\{[H_2][I_2]\} = 48.9 = (2x)^2/(1 - x)(1 - x) = (2x)^2/(1 - x)^2$.
8. In this example, if you assume $x \ll 1 \Rightarrow K_c = 4x^2 = 48.9$. Hence, $x^2 = 12.225$ and $x = 3.496$. Since $x > 1$, the assumption made is not valid (as expected since $K_c \gg 1$, *i.e.* the equilibrium lies towards the products), and a quadratic equation must be solved. Returning to the original expression of K_c in step 7: $(2x)^2/(1 - x)^2 = 48.9 \ldots (\dagger)$.
   Therefore: $4x^2 = 48.9(x^2 + 1 - 2x) = 48.9x^2 - 97.8x + 48.9$
   $\Rightarrow 44.9x^2 - 97.8x + 48.9 = 0$
   This equation is in the form $ax^2 + bx + c = 0$, with solution,

$$x = \frac{-b \pm \sqrt{b^2 - 4ac}}{2a}$$

where $a = 44.9$, $b = -97.8$ and $c = 48.9$.

$$\text{Hence: } x = \frac{97.8 \pm \sqrt{(-97.8)^2 - (4 \times 44.9 \times 48.9)}}{2 \times 44.9}$$

$= (97.8 \pm 27.971)/89.9$ This generates two values of x: 0.777 and 1.399. Since x must be less than 1, the correct answer is 0.777.

9. Determine the equilibrium concentrations:
   $[H_2] = 1 - x = 1 - 0.777 = 0.223$ M; $[I_2] = 0.223$ M; $[HI] = 2x = 2 \times 0.777 = 1.554$ M.
10. One final note. In this question, the assumption, x must be $\ll 1$ was invalid, and a quadratic equation was used. However, on close examination of (†), if the square root is taken on both sides of the equation, a linear equation is obtained, from which x can be evaluated. The expressions generated should always be closely examined for potential short-cuts in such questions!
11. $K_p = p(HI)^2/\{p(H_2)p(I_2)\} = 48.9$.

RELATIONSHIP BETWEEN ΔG AND K_p

In Chapter 3, $\Delta G°$, the change in standard Gibbs free energy, was related to both the change in the standard enthalpy and the change in the standard entropy:

$$i.e.\ \Delta G° = \Delta H° - T\Delta S°$$

ΔG is also related to K_p from the expression:

$$\boxed{\Delta G = \Delta G° + RT \ln K_p}$$

At equilibrium, $\Delta G = 0$

$i.e.\ \Delta G° + RT \ln K_p = 0$

$$RT \ln K_p = -\Delta G° \Rightarrow \ln K_p = (-\Delta G°)/(RT).$$

Hence, if $\Delta H°$ and $\Delta S°$ can be calculated for a reaction, $\Delta G°$ can be determined from the equation above, and ultimately a value for K_p, the equilibrium constant, can be obtained.

LE CHÂTELIER'S PRINCIPLE

Le Châtelier studied the influence of temperature, pressure and concentration on systems at equilibrium.

Le Châtelier's Principle states that when a system at equilibrium is disturbed, the system will counteract as far as possible the effect of the disturbance on the system.

Changes in Temperature

Consider the following two reactions:

(a) $N_{2(g)} + 3H_{2(g)} \rightleftharpoons 2NH_{3(g)}$,
$K_p = p(NH_3)^2/\{p(N_2)^1 p(H_2)^3\}$.

This is an exothermic reaction, *i.e.* ΔH° is $-$ve.

$$\Delta G = \Delta G^\circ + RT \ln K_p$$

At equilibrium, $\Delta G = 0$
$\Rightarrow \ln K_p = (-\Delta G^\circ)/(RT) = (-\Delta H^\circ/RT) + (T\Delta S^\circ/RT)$, since $\Delta G^\circ = \Delta H^\circ - T\Delta S^\circ$.
Hence, cancelling T in the second term, the expression rearranges to:

$$\ln K_p = (-\Delta H^\circ/RT) + (\Delta S^\circ/R).$$

Therefore, if the temperature is increased, the $(-\Delta H^\circ/RT)$ term is decreased (since ΔH° is $-$ve for an exothermic reaction), and so $K_p = p(NH_3)^2/\{p(N_2)^1 p(H_2)^3\}$ is decreased, *i.e.* if the temperature is increased, the system can absorb heat by the dissociation of $NH_{3(g)}$ into $N_{2(g)}$ and $H_{2(g)}$. Hence, K_p will be decreased, *i.e.* the reaction shifts in an endothermic direction to the left, as predicted by Le Châtelier's Principle. Similarly, for the above reaction, if the temperature is decreased, K_p is increased and the equilibrium shifts to the right.

(b) $N_{2(g)} + O_{2(g)} \rightleftharpoons 2NO_{(g)}$,
$K_p = p(NO)^2/\{p(N_2)p(O_2)\}$

This is an endothermic reaction, *i.e.* ΔH° is $+$ve.
At equilibrium, $\ln K_p = (-\Delta H^\circ/RT) + (\Delta S^\circ/R)$, as above.

Therefore, if the temperature is increased, the $(-\Delta H^\circ/RT)$ term is increased (made less negative, since ΔH° is $+$ve), and so $K_p = p(NO)^2/\{p(N_2)p(O_2)\}$ is increased. The equilibrium then shifts to the right. If T is decreased the equilibrium shifts to the left.

Changes in Pressure

Consider the reaction

$$N_{2(g)} + 3H_{2(g)} \rightleftharpoons 2NH_{3(g)},$$
$$K_p = p(NH_3)^2/\{p(N_2)^1 p(H_2)^3\}.$$

If the pressure of an equilibrium mixture of $N_{2(g)}$, $H_{2(g)}$ and $NH_{3(g)}$ is increased, there is a shift in the position of equilibrium in the direction that tends to reduce the pressure as predicted by Le Châtelier. From the equation of state of an ideal gas, $pV = \Delta nRT$, *i.e.* $p = (\Delta n/V)RT$. Therefore $p \propto \Delta n$ For a reduction in pressure to occur, n must decrease. Therefore, the total number of molecules must decrease. This is done by shifting the position of equilibrium from left to right, *i.e.* four gaseous molecules to two gaseous molecules.

It must be emphasised here that although the position of equilibrium shifts to the right, the value of the equilibrium constant does not change. Conversely, if the pressure is decreased, the equilibrium shifts to the left. In the case of changes in temperature, the value of K_c does change.

Changes in Concentration

$BiCl_{3(aq)}$ is a cloudy solution due to a hydrolysis reaction (reaction with water): $BiCl_{3(aq)} + H_2O_{(l)} \rightleftharpoons BiOCl_{(s)} + 2HCl_{(aq)}$. If some concentrated hydrochloric acid is added, the position of equilibrium shifts in the direction that will absorb the acid, *i.e.* from right to left. Therefore the hydrolysis reaction is considerably decreased resulting in the formation of a clear solution.

However, the solution does not absorb all the acid.

Effect of a Catalyst on Equilibrium

A catalyst is a reagent which accelerates or retards (anti-catalyst) the rate of a chemical reaction, but is not itself consumed in the reaction, and it has no effect on the equilibrium concentration or the value of the equilibrium constant.

An iron catalyst is used in the Haber process, used to manufacture ammonia according to the equation: $N_{2(g)} + 3H_{2(g)} \rightleftharpoons 2NH_{3(g)}$, $K_p = p(NH_3)^2/\{p(N_2)^1 p(H_2)^3\}$. Here the role of the catalyst is to make the reaction attain equilibrium more rapidly at the relatively low temperature employed (400–600 °C).

SUMMARY

The most important feature of this chapter is the working method for solving simple equilibrium type problems. Two important equations should be memorised: $K_p = K_c(RT)^{\Delta\nu_g}$ and $\Delta G = \Delta G^\circ + RT\ln K_p$, where at equilibrium $\Delta G = 0$.

Chapter 5

Equilibrium II: Aqueous Solution Equilibria

ACIDS AND BASES

A Brønsted–Lowry ***acid*** is a ***proton donor*** (H^+). Ionic dissociation is the breaking up of a reactant to form a cation and an anion. There are two types of acid. A ***strong acid*** is an acid which ***dissociates completely*** (100%) in solution, *i.e.* **HA** + H_2O, → H_3O^+ + A^-, where H_3O^+ represents the hydronium cation (simply water with an added proton). An example of a strong acid is **HNO_3**, since $HNO_3 + H_2O \rightarrow H_3O^+ + NO_3^-$. Table 5.1 provides other examples of strong acids. Since strong acids are dissociated completely in solution, the reverse reaction does not occur and hence an equilibrium is not established. Therefore, a forward arrow is used to illustrate the reaction. A ***weak acid*** is an acid which does not dissociate completely in solution, and hence has an equilibrium condition, *i.e.* **HA** + $H_2O \rightleftharpoons H_3O^+ + A^-$, where $K_a = \{[H_3O^+][A^-]\}/[HA]$, since the activity, a, of a pure liquid (water) is unity, as described in Chapter 4. K_a represents the equilibrium constant for the dissociation of an acid. Examples of weak acids are organic acids containing the carboxylic acid functional group RCO_2**H** (R = alkyl group), *e.g.* CH_3CO_2**H** + $H_2O \rightleftharpoons CH_3CO_2^- + H_3O^+$, where $K_a = \{[CH_3CO_2^-][H_3O^+]\}/[CH_3CO_2H]$.

A ***base*** is defined as a proton acceptor or a producer of OH^- ions. There are two types of base. ***Strong bases***, such as NaOH, KOH *etc.*, dissociate completely in solution, according to the reaction $MOH \rightarrow M^+ + OH^-$. ***Weak bases***, such as ammonia and the amines, *e.g.* NH_3, CH_3NH_2, do not dissociate 100% completely and so they exist in, equilibrium, *e.g.* $CH_3NH_2 + H_2O \rightleftharpoons CH_3NH_3^+ + OH^-$, where $K_b = \{[CH_3NH_3^+][OH^-]\}/[CH_3NH_2]$ is the equilibrium constant for base dissociation.

Table 5.1 summarises some common strong and weak acids and bases, which need to be identified in problems (R = an alkyl group *e.g.* CH_3, CH_3CH_2 *etc.*)

Table 5.1 *Some examples of strong and weak acids and bases*

Strong acids:	1. HCl	2. HNO_3	3. $HClO_4$	4. H_2SO_4
Weak acids:	1. HNO_2	2. $HClO_2$	3. *Carboxylic acids*: RCO_2H	
Strong bases:	1. NaOH	2. KOH		
Weak bases:	1. NH_3	2. *Amines*: 1° (RNH_2), 2° (R_2NH), 3° (R_3N)		

COMMON ION EFFECT

A ***common ion*** is an ion (charged species) common to two substances in the same mixture, *e.g.* in a solution of ethanoic acid, CH_3CO_2H and sodium ethanoate, CH_3CO_2Na, the common ion is the ethanoate anion, $CH_3CO_2^-$. The common ion effect occurs when the presence of extra (common) ions in the solution represses or restrains the dissociation of a species. To explain this, consider the following example:

> ***Example:*** A solution is prepared by adding 0.6 moles of sodium ethanoate, CH_3CO_2Na, and 0.8 moles of ethanoic acid, CH_3CO_2H, to water, to make up 1 dm^3. Determine the concentration of all solute species, given that $K_a(CH_3CO_2H) = 1.8 \times 10^{-5}$ at 25 °C.

Solution: To solve this problem, the working method of Chapter 4 is applied.

1. Read the question carefully—K_a problem!
2. Species present in solution: $CH_3CO_2Na_{(aq)}$, $CH_3CO_2H_{(aq)}$ and the corresponding ions (step 3 below).
3. (a) $CH_3CO_2Na_{(aq)}$ is a salt and undergoes 100% dissociation into its anion and cation, $CH_3CO_2Na_{(aq)} \rightarrow Na^+_{(aq)} + CH_3CO^-_{2(aq)}$, *i.e. the equilibrium lies completely to the ionic products!*
(b) $CH_3CO_2H_{(aq)}$ is a weak acid. Weak acids do not dissociate completely into anions and cations. Hence, the reaction is at equilibrium, *i.e.* $CH_3CO_2H + H_2O \rightleftharpoons CH_3CO_2^- + H_3O^+$.

In this solution, both reactions (a) and (b) occur simultaneously, and so both must be considered when calculating the concentration of $CH_3CO_2^-$, since each reaction acts as a source of $CH_3CO_2^-$ ions.

4. (a)

	$CH_3CO_2Na_{(aq)}$	$\rightarrow$	$Na^+_{(aq)}$	$+$	$CH_3CO^-_{2\,(aq)}$
Initial	0.6		0		0
Final	0		0.6		0.6

At the end of the reaction, nothing remains of the salt, since 100% dissociation of CH_3CO_2Na has occurred! Therefore, from the dissociation of sodium ethanoate, 0.6 mol of ethanoate anion, $CH_3CO_2^-$, is generated.

(b)	CH_3CO_2H	$+$	H_2O	$\rightleftharpoons$	$CH_3CO_2^-$	$+$	H_3O^+
Initial	0.8				0		0
Change	$-x$				$+x$		$+x$
Final	$0.8-x$				x		x

But the dissociation of the salt produces 0.6 mol of $CH_3CO_2^-$. *Hence the actual equilibrium concentration of* $CH_3CO_2^-$ *is not* x, *but* $(0.6 + x)$*!*

i.e.	CH_3CO_2H	$+$	H_2O	$\rightleftharpoons$	$CH_3CO_2^-$	$+$	H_3O^+
Final	$0.8-x$				$0.6 + x$		x

where $K_a = \{[CH_3CO_2^-][H_3O^+]\}/[CH_3CO_2H]$, since the activity of H_2O, $a = 1$.

5. Hence, $K_a = \{(0.6 + x)(x)\}/(0.8-x) = 1.8 \times 10^{-5}$.
6. Assume that $x \ll 0.8$. Hence, K_a becomes $(0.6x)/(0.8) = 1.8 \times 10^{-5} \Rightarrow x = 0.000024$.
7. $(0.000024/0.6)\% = 0.004\%$ and $(0.000024/0.8) = 0.003\%$. Since both are $<5\%$, the assumption made was valid, and hence a quadratic equation need not be solved in this particular case. Therefore $[CH_3CO_2]^- = 0.600024$ M; $[H_3O^+] = 0.000024$ M and $[CH_3CO_2H] = 0.799976$ M. Hence, % acid dissociated = $(x/0.8)\% = 0.003\%$.

   If the common ion effect had not occurred *i.e.* if there was no salt, CH_3CO_2Na present in solution, the equilibrium expression would be the following:

	CH_3CO_2H	$+$	H_2O	$\rightleftharpoons$	$CH_3CO_2^-$	$+$	H_3O^+
Initial	0.8				0		0
Change	$-x$				$+x$		$+x$
Final	$0.8-x$				x		x

Again assuming that $x \ll 0.8$, $K_a = x^2/0.8 = 1.8 \times 10^{-5}$, *i.e.* x

= 0.00379 M = [CH_3CO_2H], and the % acid dissociation = (0.00379/0.8)% = 0.47%.

Therefore, the presence of the extra source of $CH_3CO_2^-$ ions repressed the dissociation of ethanoic acid from 0.47% to 0.003%. This can be explained by Le Châtelier's Principle, as discussed in Chapter 4. The addition of ethanoate ions causes the equilibrium to shift to the left, increasing the concentration of ethanoic acid, CH_3CO_2H, and reducing the concentration of H_3O^+, *i.e.* increasing the pH!

i.e. $$CH_3CO_2H + H_2O \rightleftharpoons CH_3CO_2^- + H_3O^+$$

DISSOCIATION OF H_2O AND pH

Dissociation of H_2O

Water undergoes self-dissociation generating H_3O^+ and OH^- in low concentration, according to the equilibrium reaction: $H_2O + H_2O \rightleftharpoons H_3O^+ + OH^-$, where $K_c = \{[H_3O^+][OH^-]\}/\{[H_2O][H_2O]\}$. But, since the activity, a, of pure water is unity, the equilibrium can be described by $K_w = [H_3O^+][OH^-]$, where K_w has a value of 10^{-14} and is termed ***the dissociation constant of water***.

pH

The pH (the power of the hydronium or hydrogen ion concentration) is defined as the log to the base 10 of the hydronium ion concentration and is a means of expressing the acidity or basicity of a solution:

$pH = -\log_{10}[H_3O^+]$ or $pH = -\log_{10}[H^+]$
Similarly, an expression for the OH^- ion can be defined as
$pOH = -\log_{10}[OH^-]$
and remembered by definition as: pH + pOH = 14.

For example, the pH of a 0.15 M solution of **HNO_3** $= -\log_{10}[H_3O^+] = -\log_{10}[0.15] = 0.82$. Likewise, the pH of a 0.001 M solution of NaOH = 14−pOH = 14−(−$\log_{10}[OH^-]$) = 14−3 = 11.

The pH scale is a scale ranging from 0 to 14, with pure deionised water having an intermediate value of 7.0. The scale is shown in Figure 5.1. It is not necessary to remember the exact values, just the relative positions on the scale of both strong and weak acids and bases respectively.

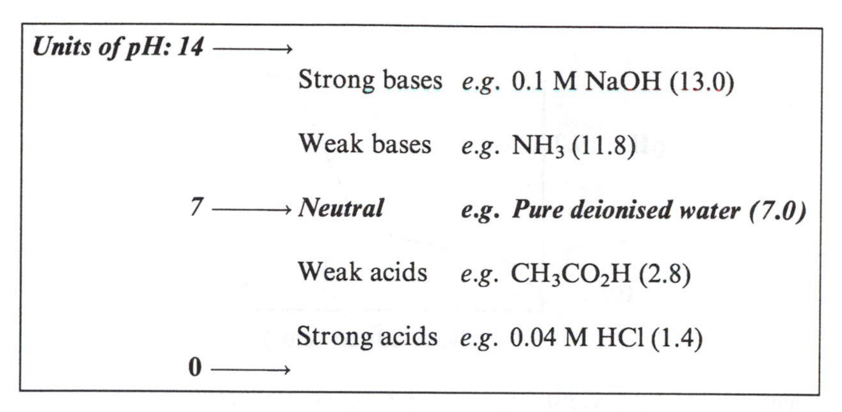

Figure 5.1 *pH scale*

Acid/Base Titrations and Indicators

As stated in Chapter 1, an acid combines with a base to form a salt and water, *i.e.* acid + base → salt + water. The ***end-point*** of a titration is the point at which equal numbers of reactive species of acid (H^+) and base (OH^-) have been mixed. This is indicated by the colour change of an ***indicator*** added to the solution. Acid–base indicators are either complex weak organic acids, HIn, or complex weak organic bases, InOH. The pH of a solution can be plotted in the form of a titration curve, which shows the pH of a solution as a function of the volume of titrant added. The correct indicator needs to be chosen for a specific titration. There are three types of titration curves, shown in Figure 5.2(a–c).

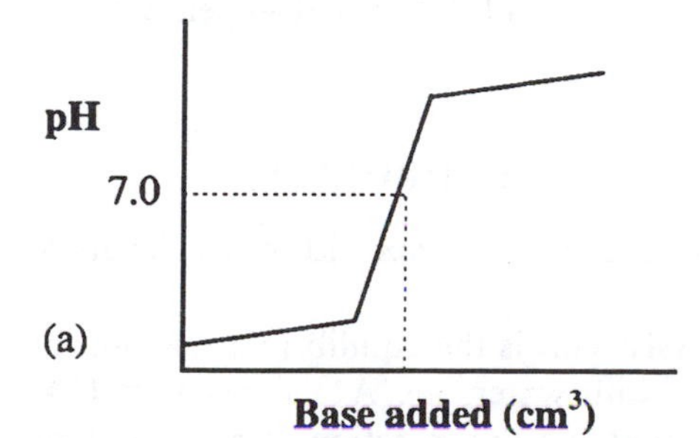

Figure 5.2(a) *Strong acid/strong base, e.g. HCl/NaOH*

An indicator which shows a colour change in the region 4–10 is needed, *e.g.* PHENOLPHTHALEIN (colourless → red).

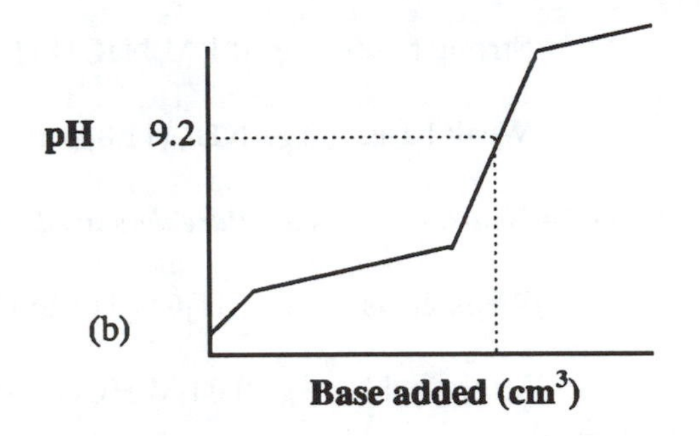

Figure 5.2(b) *Weak acid/strong base, e.g.* $CH_3CO_2H/NaOH$

Suitable indicator: PHENOLPHTHALEIN (colourless → red).

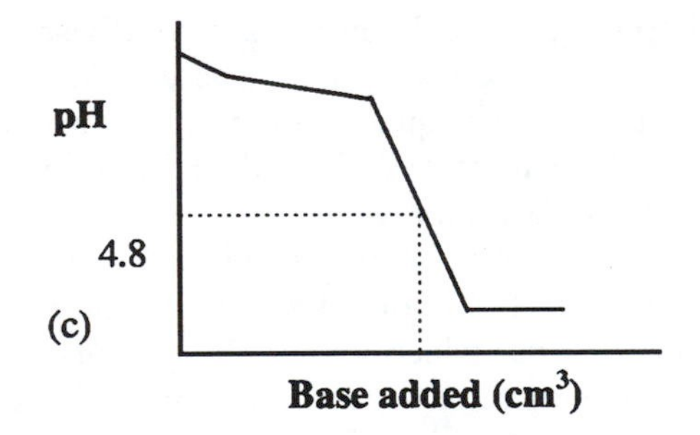

Figure 5.3(c) *Strong acid/weak base, e.g.* HCl/NH_3

Suitable indicator: METHYL RED (red → yellow).

HYDROLYSIS

Hydrolysis is the reaction of an ***ion*** with water. There are two types of hydrolysis:

(a) Anion hydrolysis: This is the equilibrium reaction of the *anion of a weak acid* (A^-) with water, *i.e.* $A^- + H_2O \rightleftharpoons HA + OH^-$. The equilibrium constant for this reaction is expressed by $K_h = \{[HA][OH^-]\}/[A^-]$, since the activity, *a*, of pure water is unity. An example of such a hydrolysis is: $CH_3CO_2^- + H_2O \rightleftharpoons CH_3CO_2H + OH^-$, with $K_h = \{[CH_3CO_2H][OH^-]\}/[CH_3CO_2{}^-]$. $CH_3CO_2^-$ hydrolyses since it is the anion of a weak acid, CH_3CO_2H.

In particular: $K_w = K_h K_a$ (remember the order: **'WHAle'!)**

(b) Cation hydrolysis: $CH^+ + H_2O \rightleftharpoons C + H_3O^+$. The equilibrium constant is expressed by $K_h = \{[C][H_3O^+]\}[CH^+]$, since the activity, *a*, of pure water is unity. An example of such a hydrolysis is: $NH_4^+ + H_2O \rightleftharpoons NH_3 + H_3O^+$, with $K_h = \{[NH_3][H_3O^+]\}/[NH_4^+]$. Only the conjugate acids of weak bases (such as NH_3) undergo cation hydrolysis.

In particular: $K_w = K_b K_h$

BUFFER SOLUTIONS

A buffer solution is a solution with an approximately ***constant pH***. A buffer contains an acid and its conjugate base in similar concentrations. Such a solution changes pH only slightly when H_3O^+ or OH^- is added. Therefore, buffers are used when the pH has to be maintained within certain restricted limits.

In the reaction $CH_3CO_2H + H_2O \rightleftharpoons CH_3CO_2^- + H_3O^+$, if both the acid and anion are present in equal concentration, the equilibrium can shift in either direction, to the right- (RHS) or to the left-hand (LHS) side:

(a) If H_3O^+ is added, the equilibrium shifts to the LHS to consume H_3O^+.
(b) If OH^- is added, the equilibrium shifts to the RHS as H_3O^+ is removed by OH^- ion, in accordance with Le Chatelier's Principle.

For either case, the concentration of the buffer remains the same. Therefore, the pH does not change to any great extent, remaining practically constant.

How to Solve a Buffer Problem on Equilibrium

Buffer problems are solved using the standard Working Method described in Chapter 4.

Example: Given that $K_a(1)$ for the first ionisation of $H_3PO_4 = 7.5 \times 10^{-3}$, $H_3PO_4 + H_2O \rightleftharpoons H_3O^+ + H_2PO_4^-$, determine the pH of the buffer solution that contains equal volumes of 0.35 M H_3PO_4, and 0.25 M NaH_2PO_4 respectively.

Solution:

1. Read the question carefully—notice that the solution is a buffer $\Rightarrow$ constant pH.
2. Identify the species present: $H_3PO_{4(aq)}$ —acid; $NaH_2PO_{4(aq)}$—salt.
3. (a) $H_3PO_4 + H_2O \rightleftharpoons H_3O^+ + H_2PO_4^-$;
   (b) $NaH_2PO_4 \rightarrow Na^+ + H_2PO_4^-$.
4. | 0.35 M | + | 0.25 M | → | **Buffer: 2 dm^3** |
   |---|---|---|---|---|
   | **0.35 mol in 1 dm^3 of H_3PO_4** | | **0.25 mol in 1 dm^3 of NaH_2PO_4** | | |

   Therefore in the buffer solution:
   0.35 moles of H_3PO_4 in 2 dm^3 $\Rightarrow$ 0.175 moles H_3PO_4 in 1 dm^3, *i.e.* 0.175 M;
   0.25 moles of NaH_2PO_4 in 2 dm^3 $\Rightarrow$ 0.125 moles NaH_2PO_4 in 1 dm^3, *i.e.* 0.125 M.
5. | | H_3PO_4 | + H_2O | $\rightleftharpoons$ | H_3O^+ | + $H_2PO_4^-$ |
   |---|---|---|---|---|---|
   | ***Initial*** | 0.175 | | | 0 | 0.125 |
   | ***Change*** | $-x$ | | | $+x$ | $+x$ |
   | ***Final*** | $(0.175 - x)$ | | | x | $(0.125 + x)$ |

6. $K_a = \{[H_2PO_4^-][H_3O^+]\}/[H_3PO_4] = \{(0.125 + x)(x)\}/(0.175 - x) = 7.5 \times 10^{-3}$ ..... (†).
7. Assume $x \ll 0.175$ and $x \ll 0.125$.
8. $K_a = (0.125\,x)/(0.175) = 7.5 \times 10^{-3} \Rightarrow x = 0.0105$ M.
9. $(0.0105/0.125)\% = 8.4\%$ and $(0.0105/0.175)\% = 6\%$. Since both are $> 5\%$, the assumptions made were invalid, and a quadratic equation (†), must now be solved:
   $x^2 + (0.125)x = (-7.5 \times 10^{-3})x + 0.0013125$
   $\Rightarrow x^2 + (0.1325)x - 0.0013125 = 0$.

   $a = 1;\ b = 0.1325;\ c = -0.0013125$, with solution

   $$x = \frac{-b \pm \sqrt{b^2 - 4ac}}{2a}$$

   $\Rightarrow x = 0.009257$ or $x = -0.141759$. The latter negative value of x is meaningless. Therefore $x = [H_3O]^+ = 0.009257$ M.
10. pH $= -\log_{10}[H_3O^+] = -\log_{10}(0.0092587) = 2.033$.

Answer: pH = 2.03

ADVANCED EQUILIBRIUM PROBLEMS— WORKING METHOD

In the previous section, buffer type problems were introduced, where the equilibrium expression involved K_a, the dissociation constant of an acid:

i.e. $HA + H_2O \rightleftharpoons H_3O^+ + A^-$, $K_a = \{[H_3O^+][A^-]\}/[HA]$,

from which the pH of the buffer solution could be determined, using the equation pH $= -\log_{10}[H_3O^+]$. The following working method deals with acid/base equilibrium type problems, where it has to be determined whether (a) acid or base dissociation (K_a or K_b) occurs, or (b) anion or cation hydrolysis (K_h) occurs. Such problems are more difficult and need to be divided into a series of steps:

1. Read the question carefully.
2. Identify the type of reaction, *i.e.* acid/base.
3. Identify the type of reactants:

1. Strong acids:	1. HCl	2. HNO_3	3. $HClO_4$	4. H_2SO_4
2. Weak acids:	1. HNO_2	2. $HClO_2$	3. *Carboxylic acids*: RCO_2H	
3. Strong bases:	1. NaOH	2. KOH		
4. Weak bases:	1. NH_3	2. *Amines*: 1° (RNH_2), 2° (R_2NH), 3° (R_3N)		

4. Write a balanced equation for the reaction. A Brønsted–Lowry acid (H^+A^-) is a proton donor (H^+) and a Brønsted–Lowry base (B) is a proton acceptor:

i.e.	HA	+	B	→	BH^+	+	A^-
	Acid (donates H^+)		Base (accepts H^+)		Cation		Anion

5. Determine the amount of each reactant (expressed in moles) present initially (the amount of product is zero at this stage).

Amount (in moles) = [Volume (in cm^3) × Molarity (in M)]/1000 (*)

6. Determine the amount (in moles) of reactants and products at the end of the reaction.
7. Calculate the concentration (*i.e.* molarity) of all species left in solution, from (*), *i.e.* Molarity (in M) = [Amount (in moles) × 1000]/Volume (in cm^3)]. *However, the volume now involved is the total volume!*
8. Determine next which of these species has the greatest concentration. *This then undergoes the equilibrium reaction:*

(a) If the highest concentration that remains is that of an acid or base, acid or base dissociation occurs:

Acid dissociation: $HA + H_2O \rightleftharpoons H_3O^+ + A^-$,
$K_a = \{[H_3O^+][A^-]\}/[HA]$;
Base dissociation: $B + H_2O \rightleftharpoons BH^+ + OH^-$,
$K_b = \{[BH^+][OH^-]\}/[B]$.

(b) If the highest concentration that remains is that of an anion or cation, anion or cation hydrolysis occurs:

Anion hydrolysis: $A^- + H_2O \rightleftharpoons HA + OH^-$,
$K_h = \{[HA][OH^-]\}/[A^-]$
Cation hydrolysis: $CH^+ + H_2O \rightleftharpoons C + H_3O^+$,
$K_h = \{[C][H_3O^+]\}/[CH^+]$

9. Calculate the concentration of all reactants and products in solution, letting $x = [H_3O^+]$.
10. Use the equilibrium expression to calculate the value of x: $\nu_A A + \nu_B B \rightleftharpoons \nu_C C + \nu_D D$, $K_c = \{[C]^{\nu_C}[D]^{\nu_D}\}/\{[A]^{\nu_A}[B]^{\nu_B}\}$, *i.e.* of the form, products/reactants:
(a) $K_c = K_a$ for acid dissociation; (b) $K_c = K_b$ for base dissociation; (c) $K_c = K_h$ for anion or cation hydrolysis, where:

$$K_w = K_h K_a \text{ (‘WHAle’!) and } K_w = K_b K_h$$

and $K_w = 10^{-14}$.
11. Calculate the pH of the solution: $pH = -\log_{10}[H_3O^+]$, where $pH + pOH = 14$.

Worked Examples

Example No. 1: In a titration of 35 cm^3 0.1 M CH_3CO_2H with 0.1 M NaOH, determine the pH of the solution after the addition of 15 cm^3 of base, given that $K_a(CH_3CO_2H) = 1.8 \times 10^{-5}$.

Solution:

1. Acid/base reaction.
2. NaOH—strong base; CH_3CO_2H—weak acid.

3. Acid/base reaction: $CH_3CO_2H + NaOH \rightarrow CH_3CO_2^- + H_2O + Na^+$ (Na^+ is a ***spectator ion***, *i.e.* an ion not actively involved in the equilibrium expression).
4. Amount of CH_3CO_2H (expressed in moles) = $(35 \times 0.1)/1000 = 3.5 \times 10^{-3}$; Amount of NaOH (expressed in moles) = $(15 \times 0.1)/1000 = 1.5 \times 10^{-3}$.
5.

	CH_3CO_2H	+ NaOH →	$CH_3CO_2^-$	+ H_2O	+ Na^+
Initial	3.5×10^{-3}	1.5×10^{-3}	0		0
End	$(3.5–1.5)\times10^{-3}$ $= 2.0\times10^{-3}$	0	1.5×10^{-3}		1.5×10^{-3}

6. Molarity of CH_3CO_2H = [amount (in moles) × 1000]/[total volume (in cm^3)]. The total volume is now $(35 + 15)$ cm^3 = 50 cm^3. Therefore, molarity of $CH_3CO_2H = (2 \times 10^{-3} \times 1000)/50 = 0.04$ M and the molarity of $CH_3CO_2^- = (1.5 \times 10^{-3} \times 1000)/50 = 0.03$ M.
7. Although the concentration of acid, $[CH_3CO_2H] = 0.04$ M, is only just greater than the concentration of anion, $[CH_3CO_2^-] = 0.03$ M, *acid dissociation is the dominant reaction which occurs!*
8. $CH_3CO_2H + H_2O \rightleftharpoons CH_3CO_2^- + H_3O^+$. This is the equilibrium reaction, where $K_a = \{[H_3O^+][CH_3CO_2^-]\}/[CH_3CO_2H]$.
9.

	CH_3CO_2H	+ H_2O ⇌	$CH_3CO_2^-$	+ H_3O^+
Initial	0.04		0.03	0
Change	$-x$		$+x$	$+x$
Final	$(0.04-x)$		$(0.03 + x)$	x

10. $K_a = \{[CH_3CO_2^-][H_3O^+]\}/[CH_3CO_2H] = \{(0.03 + x)(x)\}/(0.04-x) = 1.8 \times 10^{-5}$. Assume $x \ll 0.04 \Rightarrow (0.03x)\}/0.04 = 1.8 \times 10^{-5}$, *i.e.* $x = 2.4 \times 10^{-5}$.
    Therefore, $(2.4 \times 10^{-5})/(0.03)\% = 0.08\%$ and $(2.4 \times 10^{-5})/(0.04)\% = 0.06\%$. Both are < 5%, hence the assumption made was justified.
11. $[H_3O^+] = x = 2.4 \times 10^{-5}$ M. $pH = -\log_{10}[H_3O^+] = -\log_{10}(2.4 \times 10^{-5}) = 4.62$.

Answer: pH = 4.62

Example No. 2: In a titration of 15 cm^3 of 0.35 M $(CH_3)_3N$ with 0.25 M HCl, what is the pH of the solution after the addition of 15 cm^3 of 0.25 M HCl, given that $K_b(CH_3)_3N = 7.4 \times 10^{-5}$?

Solution:

1. Acid/base reaction.
2. $(CH_3)_3N$—weak base; HCl—strong acid.
3. Acid/base reaction: $(CH_3)_3N + HCl \rightarrow (CH_3)_3NH^+ + Cl^-$.
4. Amount of $(CH_3)_3N$ (expressed in moles) = $(15 \times 0.35)/1000 = 5.25 \times 10^{-3}$; amount of HCl (in moles) = $(15 \times 0.25)/1000 = 3.75 \times 10^{-3}$.

5.

	$(CH_3)_3N$	+ HCl	$\rightarrow$ $(CH_3)_3NH^+$	+ Cl^-
Initial	5.25×10^{-3}	3.75×10^{-3}	0	0
End	$(5.25-3.75) \times 10^{-3} = 1.5 \times 10^{-3}$	0	3.75×10^{-3}	3.75×10^{-3}

6. Molarity of $(CH_3)_3N$ = [amount of $(CH_3)_3N$ (in moles) $\times$ 1000]/total volume (in cm^3)]. The total volume is now (15 + 15) cm^3 = 30 cm^3. Therefore, the molarity of $(CH_3)_3N = (1.5 \times 10^{-3} \times 1000)/30 = 0.05$ M. Molarity of $(CH_3)_3NH^+ = (3.75 \times 10^{-3} \times 1000)/30 = 0.125$ M.
7. $[(CH_3)_3NH^+] = 0.125$ M $> [(CH_3)_3N] = 0.05$ M. Therefore, ***cation hydrolysis occurs!***
8. $(CH_3)_3NH^+ + H_2O \rightleftharpoons (CH_3)_3N + H_3O^+$.
   This is the equilibrium reaction,
   where $K_h = \{[(CH_3)_3N][H_3O^+]\}/[(CH_3)_3NH^+]$.

9.

	$(CH_3)_3NH^+$ + H_2O	$\rightleftharpoons$ $(CH_3)_3N$	+ H_3O^+
Initial	0.125	0.05	0
Change	$-x$	$+x$	$+x$
Final	$(0.125-x)$	$(0.05 + x)$	x

10. $K_h = \{[(CH_3)_3N][H_3O^+]\}/[(CH_3)_3NH^+] = \{(0.05 + x)(x)\}/(0.125-x)$.
11. $K_w = K_b K_h \Rightarrow K_h = K_w/K_b = 10^{-14}/(7.4 \times 10^{-5}) = 1.351 \times 10^{-10}$.
12. Assume $x \ll 0.05 \Rightarrow (0.05\,x)/0.125 = 1.351 \times 10^{-10}$, *i.e.* $x = 3.378 \times 10^{-10}$ M.
    Therefore, $(3.378 \times 10^{-10})/(0.05)\% = 6.8 \times 10^{-7}$ % and also, $(3.378 \times 10^{-10})/(0.125)\% = 2.7 \times 10^{-7}$ %. Both are $<5\%$, hence the assumption made was justified.
13. $[H_3O^+] = x = 3.378 \times 10^{-10}$ M.
    $pH = -\log_{10}[H_3O^+] = -\log_{10}(3.378 \times 10^{-10}) = 9.47$.

Answer: pH = 9.47

POLYPROTIC ACIDS

Polyprotic acids are acids which *have more than one replaceable proton.* Examples include H_2SO_4 (e_A = 2), H_3PO_4 (e_A = 3), *etc.*, where e_A is the number of replaceable hydrogens. Therefore, sulfuric acid, H_2SO_4, has two dissociation constants, $K_a(1)$ and $K_a(2)$ respectively, corresponding to the following reactions:

$H_2SO_4 + H_2O \rightleftharpoons H_3O^+ + HSO_4^-$, where $K_a(1) = \{[H_3O^+][HSO_4^-]\}/[H_2SO_4] = 1.3 \times 10^{-2}$ and $HSO_4^- + H_2O \rightleftharpoons H_3O^+ + SO_4^{2-}$, where $K_a(2) = \{[H_3O^+][SO_4^{2-}]\}/[HSO_4^-] = 6.3 \times 10^{-8}$. As expected, $K_a(2) = 6.3 \times 10^{-8} < K_a(1) = 1.3 \times 10^{-2}$, since HSO_4^- is a relatively weak acid and is negatively charged.

SOLUBILITY PRODUCT

In precipitation reactions, slightly soluble products often form, and there is an equilibrium between the solid and the ions of the saturated solution, *e.g.* for the reaction $AgCl_{(s)} \rightleftharpoons Ag^+{}_{(aq)} + Cl^-{}_{(aq)}$, the forward reaction is a dissolution reaction, and the back reaction is a precipitation reaction $\Rightarrow K_c = \{[Ag^+{}_{(aq)}][Cl^-{}_{(aq)}]\}/[AgCl_{(s)}]$. But, since the activity, *a*, of a solid is unity, this means the equilibrium constant can be expressed as $K_{sp} = [Ag^+{}_{(aq)}][Cl^-{}_{(aq)}]$, where K_{sp} is termed the ***solubility product***. Likewise for the reaction $Mg(OH)_{2(s)} \rightleftharpoons Mg^{2+}{}_{(aq)} + 2OH^-{}_{(aq)}$, $K_{sp} = [Mg^{2+}][OH^-]^2$, *etc.*

Solubility Product Equilibrium Problems

Example No. 1: The solubility product of calomel, Hg_2Cl_2 is 1.1×10^{-18}. Determine the molar solubility of Hg_2Cl_2.

Solution:

1. Solubility product question!
2. $Hg_2Cl_{2(s)} \rightleftharpoons Hg_2^{2+}{}_{(aq)} + 2Cl^-{}_{(aq)}$—Heterogeneous equilibrium (2 phases).
3. $K_{sp} = [Hg_2^{2+}{}_{(aq)}][Cl^-{}_{(aq)}]^2$.
4.

	$Hg_2Cl_{2(s)}$	$\rightleftharpoons$	$Hg_2^{2+}{}_{(aq)}$	+	$2Cl^-{}_{(aq)}$
Initial conc.			0		0
Change			$+x$		$+2x$
Final conc.			$+x$		$+2x$

5. $K_{sp} = (x)(2x)^2 = 1.1 \times 10^{-18} \Rightarrow 4x^3 = 1.1 \times 10^{-18}$. Therefore, $x = 6.5 \times 10^{-7}$ M.
6. $[Hg_2^{2+}] = x = 6.5 \times 10^{-7}$ M, and $[Cl^-] = 2x = 2(6.5 \times 10^{-7}) = 1.3 \times 10^{-6}$ M.
7. From the reaction: $1[Hg_2Cl_2] \equiv 1[Hg_2^{2+}] \equiv 2[Cl^-] \Rightarrow$ molar solubility of $[Hg_2Cl_2] = 6.5 \times 10^{-7}$ M, since $1[Hg_2Cl_2] \equiv 1[Hg_2^{2+}]$.

Answer: Molar solubility of $Hg_2Cl_2 = 6.5 \times 10^{-7}M$

Example No. 2: Determine the solubility of $Zn(CN)_2$ at 298 K in the presence of 0.15 M KCN, given that K_{sp} $Zn(CN)_2$ is 8.0×10^{-12}.

Solution:

1. Solubility product question!
2. $Zn(CN)_{2(s)} \rightleftharpoons Zn^{2+}{}_{(aq)} + 2CN^-{}_{(aq)}$—heterogeneous equilibrium (two phases).
3. $K_{sp} = [Zn^{2+}{}_{(aq)}][CN^-{}_{(aq)}]^2$.
4.

	$Zn(CN)_{2(s)}$	$\rightleftharpoons$	$Zn^{2+}{}_{(aq)}$	+	$2CN^-{}_{(aq)}$
Initial conc.			0		0.15
Change			$+x$		$+2x$
Final conc.			$+x$		$(0.15 + 2x)$

5. $K_{sp} = (x)(0.15 + 2x)^2 = 8 \times 10^{-12}$. Assume $2x \ll 0.15 \Rightarrow (0.15)^2x = 8 \times 10^{-12}$, and therefore $x = 3.6 \times 10^{-10}$ M. Hence, $(3.6 \times 10^{-10})/0.15\% = 2.4 \times 10^{-7}\%$, which is $\ll 5\%$, *i.e.* the assumption made was justified.
6. $[Zn^{2+}{}_{(aq)}] = 3.6 \times 10^{-10}$ M and $[CN^-{}_{(aq)}] = (0.15 + 2x)$, *i.e.* approximately 0.15 M.
7. From the reaction, $1[Zn(CN)_{2(s)}] \equiv 1[Zn^{2+}] \equiv 2[CN^-] \Rightarrow$ solubility of $[Zn(CN)_2] = 3.6 \times 10^{-10}$ M, since $1[Zn(CN)_{2(s)}] \equiv 1[Zn^{2+}]$.

Answer: Molar solubility of $Zn(CN)_2 = 3.6 \times 10^{-10}M$

SUMMARY

Figure 5.3 summarises the five types of problems on aqueous solution equilibrium.

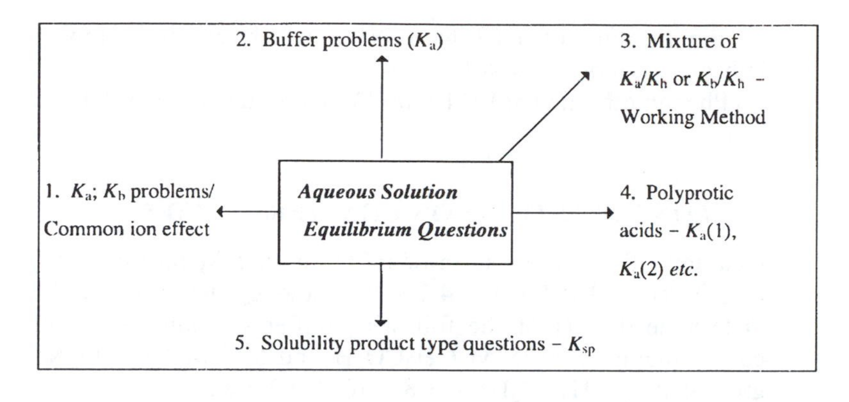

Figure 5.3 *Summary of numerical problem solving in aqueous solution type equilibrium questions*

The final two sections of this chapter contain a short multiple-choice test, and three longer, more challenging questions. The only way of becoming skilled at equilibrium type questions is to work through a series of standard problems, applying the working method.

MULTIPLE-CHOICE TEST

1. The acid HA has an acid dissociation constant, K_a. The equilibrium constant for the reaction, $A^- + H_3O^+ \rightleftharpoons HA + H_2O$, is:
   (a) K_a/K_w (b) $1/K_a$ (c) K_a (d) K_w/K_a
2. A sample of sea water was analysed in the laboratory to have a pH = 8.32. What is the concentration of OH^- ion in the sample (in M)?
   (a) 4.8×10^{-9} (b) 5.68 (c) 2.1×10^{-6} (d) −5.68
3. What is the molar solubility of MgF_2, in 0.25 M NaF solution, given that K_{sp} of $MgF_2 = 8 \times 10^{-8}$ M?
   (a) 2.6×10^{-6} (b) 5.1×10^{-6} (c) 2×10^{-4} (d) 1.3×10^{-6}
4. pK_a of HOCl = 7.46. What is the pH of a 0.35 M HOCl solution?
   (a) 3.96 (b) 0.46 (c) 0.87 (d) 7.46
5. The reaction $2Na_{(g)} \rightleftharpoons Na_{2(g)}$ has an equilibrium constant of 10^{-4} at 298 K. If 0.2 mol of $Na_{(g)}$ are introduced into a 1 dm^3 volume at that temperature, the equilibrium pressure of $Na_{(g)}$ (in bars) is ($R = 0.08314$ dm^3 bar K^{-1} mol^{-1}):
   (a) 5.0 (b) 0.3 (c) 0.03 (d) 1.8

6. In the titration of HCl with NH_3, which of the following indicators is most suitable?
(a) Phenolphthalein (b) EDTA (c) Methyl Red (d) Thymolphthalein

LONG QUESTIONS ON CHAPTERS 4 AND 5

1. Determine the concentration of HCO_3^- in 0.15 M carbonic acid, H_2CO_3, given that $K_a(1) = 4.2 \times 10^{-7}$ and $K_a(2) = 5.6 \times 10^{-11}$.
2. Determine the pH of the following buffer solution containing equal volumes of 0.35 M CH_3CO_2H and 0.15 M CH_3CO_2Na, given that $K_a(CH_3CO_2H) = 1.8 \times 10^{-5}$ at 298 K.
3. What is the pH of a solution after the addition of 30 cm^3 0.1 M $CH_3NH_{2(aq)}$ in a titration with 25 cm^3 0.1 M $HCl_{(aq)}$, given that $K_b(CH_3NH_{2(aq)}) = 4.4 \times 10^{-4}$ at 298 K.

Chapter 6

Electrochemistry I: Galvanic Cells

INTRODUCTION TO ELECTROCHEMISTRY

Energy is the ability to do work. The Law of Conservation of Energy states that energy cannot be created or destroyed, but is converted from one form to another. There are two types of energy, potential energy, which is stored energy, and kinetic energy, which is the energy a body possesses by virtue of its motion. Hence, all energy is either potential or kinetic, each of which has its own forms, as shown in

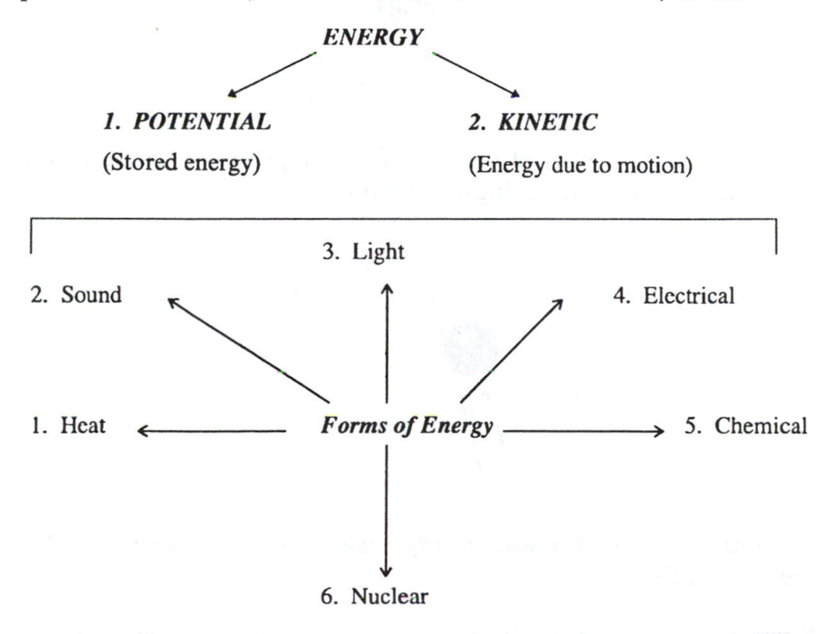

Figure 6.1 *Different forms of energy*

Figure 6.1. For a conservative system, potential energy + kinetic energy = constant.

In Chapters 6 and 7, the interconversion between two of these forms of energy, electrical and chemical energy will be considered. Electrochemistry is concerned with the chemical changes produced by an electric current and the production of electricity by a chemical reaction. There are two types of electrochemical energy conversion (Figure 6.2):

(a) Chemical → electrical energy (conversion of potential energy to kinetic energy).

This is a *downhill* process and this energy conversion occurs in the ***galvanic cell.***

(b) Electrical → chemical energy (conversion of kinetic energy to potential energy).

This is an *uphill* process and occurs in the ***electrolytic cell.*** The process is *uphill* as electrical energy has to be pumped into the system, *i.e.* in electrolytic cells, a battery is always present to provide energy.

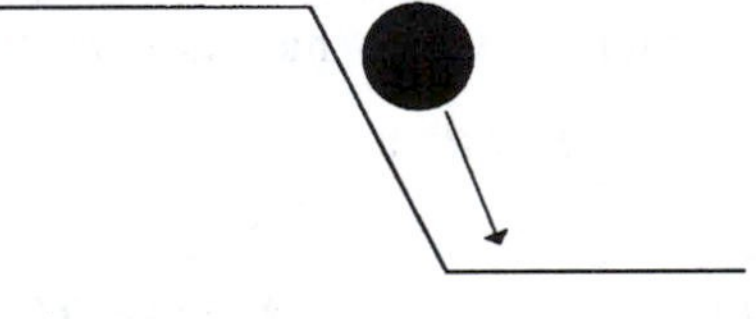

(a) Chemical → electrical energy—*downhill* process—occurs in the galvanic cell, like a ball rolling down a hill.

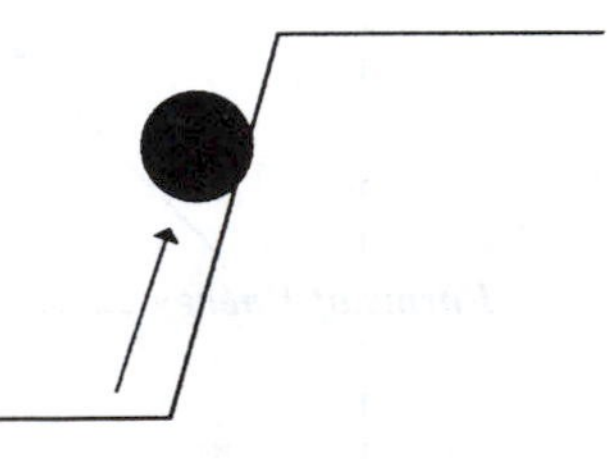

(b) *E*lectrical → chemical energy—*uphill* process—occurs in the *E*lectrolytic Cell

Figure 6.2 *Galvanic and electrolytic cells—downhill and uphill energy conversions*

In galvanic cells, an ammeter (which measures the current) or voltmeter (which measures the potential) is present to measure the flow of current, *i.e.* to measure E°_{cell}. This will be discussed later in Chapter 7.

REDOX REACTIONS—REVISION

ΔG, the Gibbs free energy change of a reaction and K, the equilibrium constant, can be determined from galvanic cell calculations. Before describing the various types of electrodes used in galvanic cells, knowledge of the fundamentals underlying oxidation–reduction reactions (redox reactions) is required. The next section summarises the fundamental concept of a redox reaction. A knowledge of redox reactions is essential in understanding the processes occurring at electrodes in both galvanic and electrolytic cells respectively.

Redox Reactions

The oxidation number of an element is the apparent charge an atom of that element has in an anion, cation, compound or complex. An ***anion*** is a negatively charged species such as Cl^-, NO_3^-, SO_3^{2-}, *etc.* A ***cation*** is a positively charged species, such as Na^+, NH_4^+, $[Mn(OH_2)_6]^{2+}$, *etc.* (An easy way to remember this is: Anion—negatively charged). Table 6.1 is a summary of the rules used to determine the oxidation number of an element.

Five examples illustrating the calculation of oxidation numbers follow:

Table 6.1 *Rules for assigning oxidation numbers*

1. The oxidation number of any atom as a free element is equal to zero, *e.g.* Na, Cl_2, Fe, S_8, He, *etc.*
2. The elements of Group 1 (***alkali metals***, s^1p^0) have an oxidation number of I ($s^1 \rightarrow s^0$) in compounds, *i.e.* Na, K, *etc.* The elements of Group 2 (***alkaline earth metals***, s^2p^0) have an oxidation number of II ($s^2 \rightarrow s^0$) in compounds, *i.e.* Be, Mg, Ca, *etc.* Group 1 and Group 2 elements all tend to lose their one or two valence electrons in order to attain the stable inert gas core configuration of [He], [Ne], *etc.*
   The elements of Group 17 (***the halogens***, s^2p^5), F, Cl, Br, I, *etc.*, normally have an oxidation number of $-$I ($s^2p^5 \rightarrow s^2p^6$) in compounds. All these elements need just one more electron to attain a stable inert gas core.
3. Oxygen, with an electron configuration of $[He]2s^22p^4$, normally has an oxidation number of $-$II, by gaining two electrons to form [Ne]. In peroxides (compounds with an O–O bond), such as H_2O_2, oxygen has an oxidation number of $-$I.

Table 6.1 *(contd)*

4. Hydrogen, $1s^1$, usually has an oxidation number of I ($s^1 \rightarrow s^0$). In metal hydrides, such as NaH, hydrogen has an oxidation number of $-$I.
5. The sum of the oxidation numbers of all the atoms is equal to the charge on the anion, cation, compound or complex. For example in the tetrafluoroborate anion, BF_4^-, the sum of the oxidation numbers must be equal to -1, whereas in the complex $[Fe(OH_2)_6]Cl_3$, the sum of the oxidation numbers must be equal to zero, *i.e.* $[B^{III}F_4^{-I}]^-$ and $[Fe^{III}(O^{-II}H_2^{I})_6]Cl_3^{-I}$

Examples of the calculation of oxidation numbers: What is the electron configuration of the free element marked with an asterisk (*), and of the element in the appropriate oxidation state in the given anion, cation, compound or complex in:

(a) $H_2S^*O_4$; **(b)** KAl^*Cl_4; **(c)** $Cr_2^*O_7^{2-}$; **(d)** $N^*H_4^+$;
(e) $[Co^*(NH_3)_6]Cl_3$

(a) S: $[Ne]3s^23p^4$; $2(I) + x + 4(-II) = 0$; $x = 6$; $S^{VI} = [Ne]s^0p^0$.
(b) Al: $[Ne]3s^23p^1$; $(I) + x + 4(-I) = 0$; $x = 3$; $Al^{III} = [Ne]s^0p^0$.
(c) Cr: $[Ar]4s^23d^4$, but is more correctly written as $[Ar]4s^13d^5$ due to the *extra stability* of the half-filled $3d$ sublevel;
$2x + 7(-II) = -2$; $x = 6$; $Cr^{VI} = [Ar]$.
(d) N: $[He]2s^22p^3$; $x + 4(I) = 1$; $x = -3$; $N^{-III} = [Ne]s^0p^0$.
(e) $[Co^*(NH_3)_6]Cl_3$: first separate into cation and anion, *i.e.* $[Co^*(NH_3)_6]^{3+}$ and $3Cl^-$ respectively. NH_3 is a neutral ligand and therefore has an oxidation number of 0; Co: $[Ar]4s^23d^7$; $x + 6(0) = 3$; $x = 3$; but electrons are removed from the $4s$ level before the $3d$, therefore $Co^{III} = [Ar]3d^6$.

Oxidation **i**s the **l**oss of electrons and **r**eduction **i**s the **g**ain of electrons ('**OILRIG**'). A more useful definition of reduction is (as its name suggests) a decrease in the oxidation number of a species, and consequently, oxidation is an increase in the oxidation number.

e.g. $Mn^{VII}O_4^- + 5e \rightarrow Mn^{II}$
VII: $[Ar]3d^0$ II: $[Ar]3d^5$
($7 \rightarrow 2$ . . . decrease, therefore reduction!)
$Fe^{II} \rightarrow Fe^{III} + e$
II: $[Ar]3d^6$ III: $[Ar]3d^5$
($2 \rightarrow 3$ . . . increase, therefore oxidation)

An ***oxidising agent*** is a species which causes some other species to be oxidised, while itself being reduced. A ***reducing agent*** is a species which causes some other species to be reduced, the reducing agent being oxidised in the process. Thus, an oxidation–reduction reaction, *i.e.* a redox reaction may be written as:

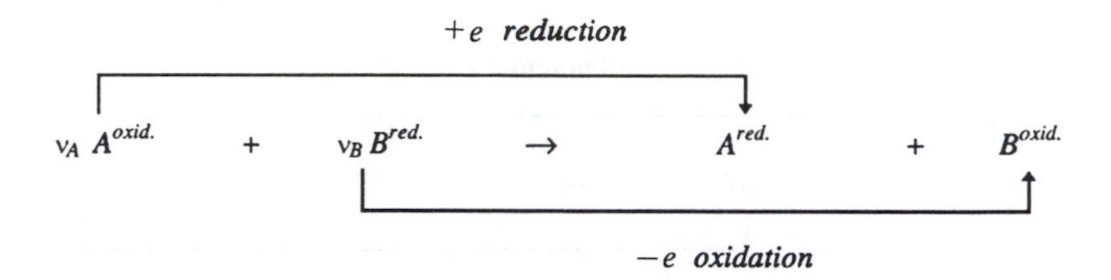

It is then a process of equating the number of electrons transferred in the two half-reactions + νe and − νe, respectively, so that the total number of electrons involved in the reduction half-reaction (e_B) is equal to the total number of electrons involved in the oxidation half-reaction (e_A):

$$\nu_A e_A = \nu_B e_B$$

where ν_A and ν_B are termed the ***stoichiometry factors***, and e_A and e_B are the number of reactive species, *i.e.* the number of electrons transferred in the two half-reactions. Consider the following example of a redox reaction:

Determine ν_A and ν_B in the following reaction:

$\nu_A\, NO_2^- + \nu_B\, MnO_4^- \rightarrow$

A. $[N^{III}O_2]^- \rightarrow [N^{V}O_3]^- + 2e$
   III V
   ($3 \rightarrow 5$ ... increase, therefore oxidation)

B. $Mn^{VII} + 5e \rightarrow Mn^{II}$
   VII II
   ($7 \rightarrow 2$ ... decrease, therefore reduction)

Hence: $\nu_A e_A = \nu_B e_B$

$\nu_A \times 2 = \nu_B \times 5$

One solution to this equation is:

$\nu_A = 5; \nu_B = 2$

meaning 5 NO_2^- react with 2 MnO_4^-

i.e. 5 $NO_2^- \equiv$ 2 MnO_4^-

i.e. 5 NO_2^- + 2 $MnO_4^- \rightarrow$

GALVANIC CELLS

A galvanic cell is composed of two half-cells, each of which is associated with the process occurring at one of the two electrodes. As in a redox reaction in inorganic chemistry, the oxidation half-reaction is not divorced from the reduction half-reaction, but both collectively act as a ***couple***.

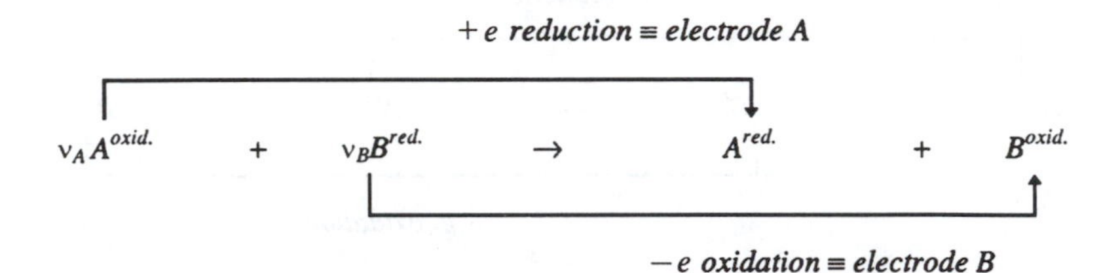

Similarly, it is impossible to isolate one half-cell from the other, nor can the electrode potential of one half-cell be measured without reference to the second half-cell.

In any electrochemical cell (both galvanic and electrolytic cells), the electrode at which reduction takes place is called the ***cathode***, and the electrode at which oxidation takes place is called the ***anode***. A useful way of remembering this is the mnemonic **'CROA'**, *i.e.* **c**athode–**r**eduction **a**node–**o**xidation.

Types of Electrodes Used in Galvanic Cells

There are four basic types of electrodes used in galvanic cells:

(a) Metal–Metal-Ion Electrode (*e.g.* Figure 6.3).

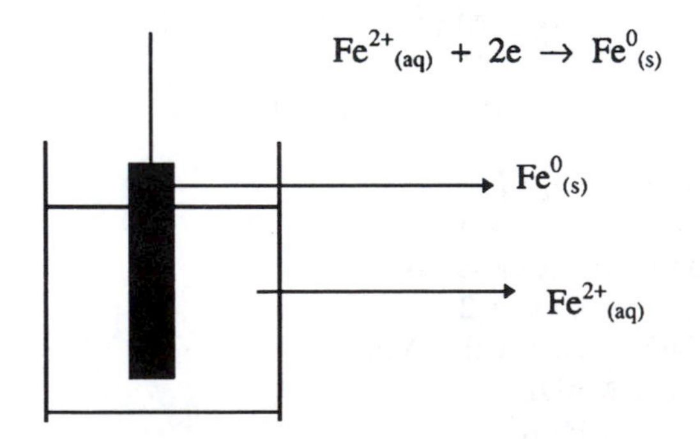

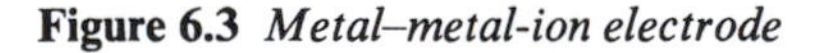
Figure 6.3 *Metal–metal-ion electrode*

A metal–metal-ion electrode is simply a bar of metal immersed in a solution of its own ions. Other examples include, $Cu_{(s)}|Cu^{2+}_{(aq)}$,

$Zn_{(s)}|Zn^{2+}{}_{(aq)}$ *etc.*, where the short vertical line represents a phase boundary or junction.

(b) Metal Ion in Two Different Valence States (or Oxidation Numbers) (*e.g.* Figure 6.4)

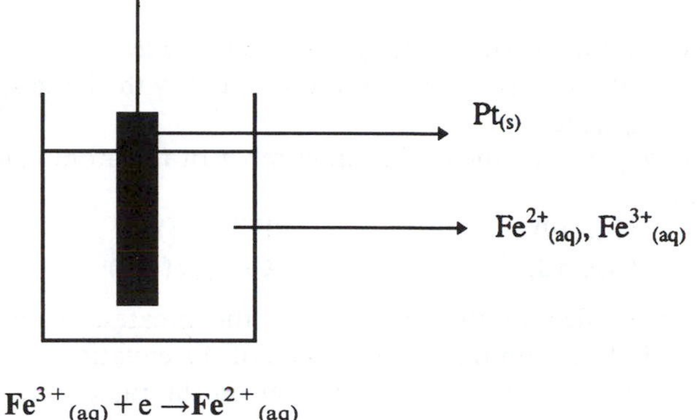

Figure 6.4 *Electrode with metal ions in different valence states*

In this case, the metal immersed in solution is an inert metal, such as platinum Pt, *i.e.* one which is not involved in the electrode process.

Another example of such an electrode is $Pt|Cr^{3+}{}_{(aq)}$, $Cr^{2+}{}_{(aq)}$.

(c) Gas–Ion Electrode (*e.g.* Figure 6.5)

Gas electrodes consist of an inert metal, such as Pt, with a gas acting as either the oxidised or reduced species, bubbled around the

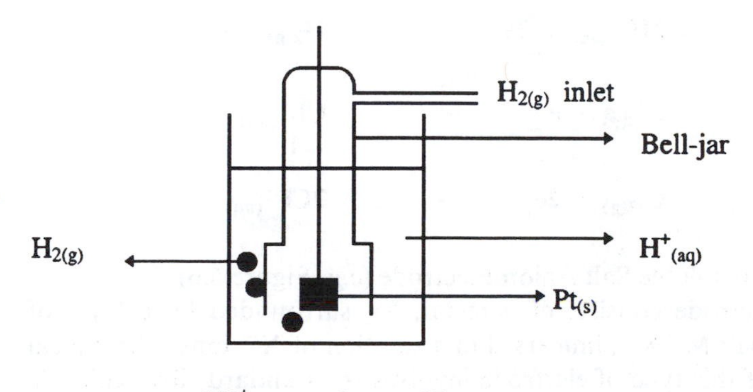

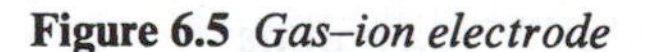

Figure 6.5 *Gas–ion electrode*

inert conductor. The ***standard hydrogen electrode*** (SHE), discussed further in the next section, is one example of a gas electrode. Hydrogen gas is bubbled around the platinum foil, covered with very finely divided platinum and immersed in a solution of H^+ ions.

In order to write the reduction half-reaction for a gas electrode, the following procedure is adopted:

1. Identify the gas involved, *e.g.* $H_{2(g)}$, $Cl_{2(g)}$, $O_{2(g)}$, *etc.*
2. Write down the corresponding ion associated with the gas, *i.e.* $H^+_{(aq)}$, $Cl^-_{(aq)}$, *etc.*
3. For each couple, state the oxidation number of the atom in each species, *i.e.*

   $H_{2(g)}$ (0) $\qquad$ $H^+_{(aq)}$ (I)
   $Cl_{2(g)}$ (0) $\qquad$ $Cl^-_{(aq)}$ (−I)

4. From step 3, identify the species with the greatest oxidation number and place it on the left-hand side of the equation and the species with the lowest oxidation number on the right-hand side to show the reduction, *i.e.*

$$\underset{I}{H^+_{(aq)}} \rightarrow \underset{0}{H_{2(g)}}$$

$$\underset{0}{Cl_{2(g)}} \rightarrow \underset{-I}{Cl^-_{(aq)}}$$

5. Balance the charges with the appropriate number of electrons in each case:

e.g.

$$\underset{I}{H^+_{(aq)}} + e \rightarrow \underset{0}{H_{2(g)}}$$

i.e.

$$\mathbf{2H^+_{(aq)} + 2e \rightarrow H_{2(g)}}$$

and

$$\underset{0}{Cl_{2(g)}} + e \rightarrow \underset{-I}{Cl^-_{(aq)}}$$

i.e.

$$\mathbf{Cl_{2(g)} + 2e \rightarrow 2Cl^-_{(aq)}}$$

(d) Metal–Insoluble Salt Anion Electrode (*e.g.* Figure 6.6)

This electrode consists of a metal, M, surrounded by a layer of insoluble salt M^+X^-, immersed in a solution of X^- ions. The typical example of this type of electrode involves the standard silver chloride precipitation reaction, *i.e.* $Ag^+_{(aq)} + Cl^-_{(aq)} \rightarrow AgCl_{(s)}$. The electrode consists of a bar of silver metal, $Ag_{(s)}$, surrounded by a layer of $AgCl_{(s)}$ salt, immersed in a solution of chloride $Cl^-_{(aq)}$ anions.

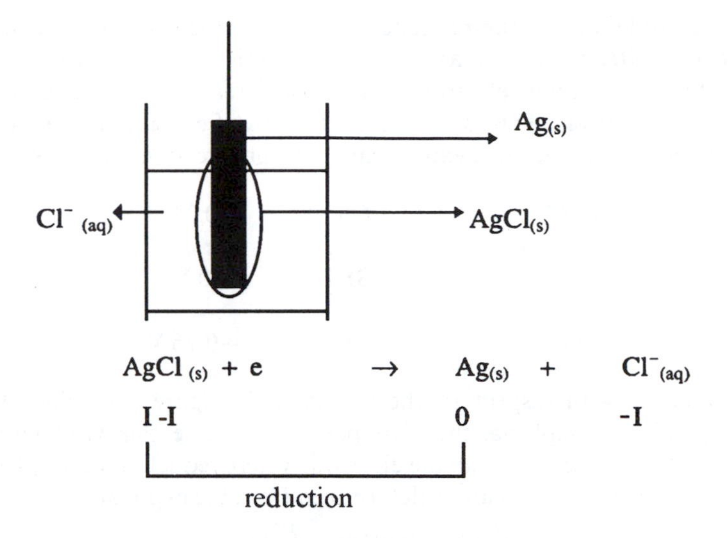

Figure 6.6 *Metal–insoluble salt anion electrode*

Notice the reduction of the silver cation, Ag^+, to metallic silver, $Ag_{(s)}$, in the above half-reaction, Ag(I) → Ag(0).

Standard EMF of a Cell

In a galvanic cell, chemical energy is converted into electrical energy (potential energy to kinetic energy), and the electricity produced is measured with an ammeter or a voltmeter. The standard electromotive force (EMF) of a cell, $E°_{cell}$, is defined as:

> $\boldsymbol{E°_{cell} = E°_{RHE} - E°_{LHE}}$, where RHE represents the right-hand electrode and LHE represents the left-hand electrode, respectively, *i.e.* the difference of the two standard electrode potentials, where **standard reduction potentials**, $E°$, are potentials measured with respect to the **standard hydrogen electrode (SHE)** at 25 °C (298 K) (*i.e.* standard state conditions = most stable state of a substance), with 1 M concentration of each ion in solution and 1 bar pressure of each gas involved.

$E°$ values are, *by convention*, always written as reduction processes, *e.g.*

$$Cu^{2+}_{(aq)} + 2e \rightarrow Cu^0_{(s)} \qquad E° = +0.34\ V;$$
$$Zn^{2+}_{(aq)} + 2e \rightarrow Zn^0_{(s)} \qquad E° = -0.76\ V.$$

The electrode used as the reference electrode is the ***standard hydrogen electrode*** (***SHE***), and is assigned a potential of zero volts. The potentials of all other electrodes are quoted relative to the standard hydrogen electrode. This is similar to the idea of considering sea-level as zero elevation, and then expressing all heights relative to this level:

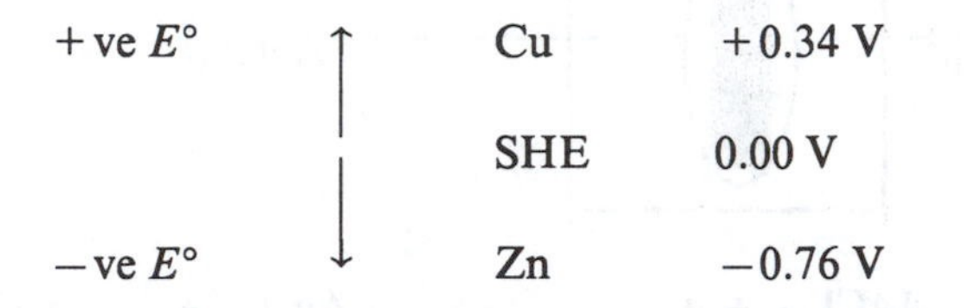

Therefore, with respect to the Daniel cell (Figure 6.7), since the $Cu^{2+}{}_{(aq)}|Cu^{0}{}_{(s)}$ couple has the more positive $E°$ value, this will then act as the cathode in the cell, *i.e.* the electrode where reduction takes place, and consequently $E°_{cell}$ can be determined from the expression

$$E°_{cell} = E°_{RHE} - E°_{LHE}$$

where the RHE has the more +ve value, *i.e.* the **cathode–reduction**, (‘**CROA**’)

$$E°_{cell} = 0.34 - (-0.76)\ V = +1.10\ V$$

This value will be registered on the voltmeter in the Daniel cell (Figure 6.7).

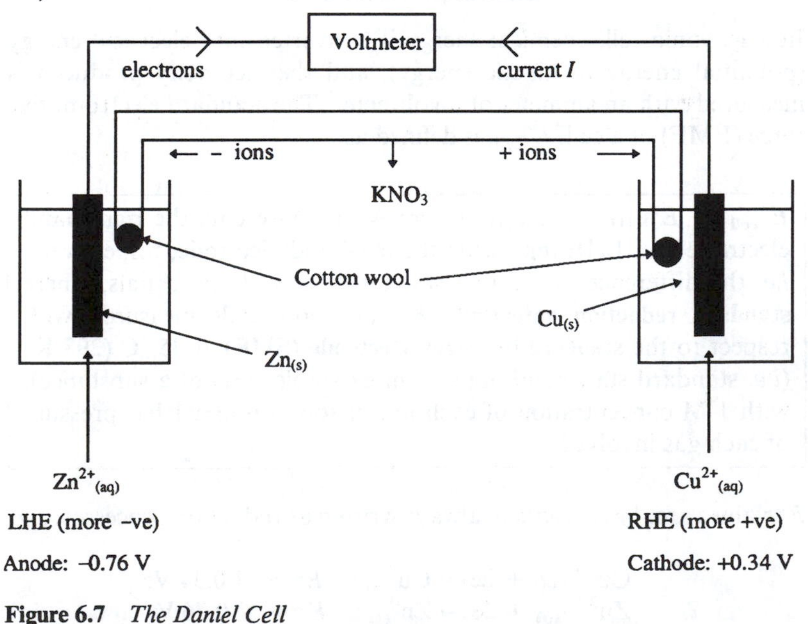

Figure 6.7 *The Daniel Cell*

Summary:

Cathode reaction: $Cu^{2+}{}_{(aq)} + 2e \rightarrow Cu^{0}{}_{(s)}$ $E° = +0.34$ V
Anode reaction: $Zn^{0}{}_{(s)} \rightarrow Zn^{2+}{}_{(aq)} + 2e$ $E° = -0.76$ V

Net cell reaction: $Cu^{2+}{}_{(aq)} + Zn^{0}{}_{(s)} \rightarrow Cu^{0}{}_{(s)} + Zn^{2+}{}_{(aq)}$
$E°_{cell} = +1.10$ V

Cell Diagrams

Galvanic cells can be represented by a shorthand notation called a cell diagram. For the Daniel cell, this one-line representation of the cell is given as:

$$Zn_{(s)}|Zn^{2+}{}_{(aq)}||Cu^{2+}{}_{(aq)}|Cu_{(s)}$$

The short vertical lines represent phase boundaries or junctions. The two vertical lines in the centre represent a device called a ***salt bridge***, which has three functions:

1. It physically separates the two electrodes, *i.e.* the cathode and the anode.
2. It provides electrical continuity within the galvanic cell, *i.e.* a path for migrating cations and anions.
3. It reduces the so-called ***liquid junction potential***. This is a voltage or potential generated when two dissimilar solutions are in contact with each other. Such a potential is produced as a result of unequal cation and anion migration across a junction. A salt bridge, as its name suggests, consists of ions (charged species), which migrate at practically equal rates. An example of such a species is the inorganic salt potassium nitrate, KNO_3, which consists of K^+ cations and $NO_3{}^-$ anions respectively.

In the one-line representation of the cell, the cathode (where reduction takes place; 'CROA') by convention is shown on the right-hand side, and the anode (where oxidation takes place) is written on the left-hand side, *i.e.* for a galvanic cell this takes the form:

Anode (Oxidation) || Cathode (Reduction)
LHE RHE

Therefore, in the Daniel cell, at the anode, metallic zinc gives up 2e to form $Zn^{2+}{}_{(aq)}$ ions. These electrons move from the anode and travel *via* the external circuit to the cathode. At the cathode, the $Cu^{2+}{}_{(aq)}$ ions combine with the 2e, producing more copper metal. This means that at the anode the zinc becomes corroded, and gradually is eaten away, whereas at the cathode copper metal is deposited. Since electrons travel from the anode to the cathode (an easy way to remember this is the two vowels, **a**node → **e**lectrons), this implies that electric current, *I* (conventional direction of flow), must travel in the opposite direction. Since oxidation occurs at the anode ('CROA'), *i.e.* a loss of electrons ('OILRIG': oxidation–loss of electrons; reduction–gain of electrons), $Zn^{2+}{}_{(aq)}$ ions are produced, and so NO_3^- anions, which are negatively charged, will migrate from the salt bridge towards the anode, to compensate this generation of positive $Zn^{2+}{}_{(aq)}$ cations. Likewise, K^+ cations will travel across the salt bridge towards the cathode.

The Electrochemical Series or the Activity Series

An important application of standard electrode potentials is in ranking substances according to their reducing and oxidising powers. The electrochemical or activity series is summarised in Table 6.2.

Table 6.2 *The Electrochemical or Activity Series*

Scale	*E°/V*	*Reduction*
−3.00		
	−3.09	$Li^+ + e \rightarrow Li$
	−2.93	$K^+ + e \rightarrow K$
	−2.71	$Na^+ + e \rightarrow Na$
	−2.36	$Mg^{2+} + 2e \rightarrow Mg$
−2.00		
	−1.66	$Al^{3+} + 3e \rightarrow Al$
	−1.18	$Mn^{2+} + 2e \rightarrow Mn$
−1.00		
	−0.76	$Zn^{2+} + 2e \rightarrow Zn$
	−0.44	$Fe^{2+} + 2e \rightarrow Fe$
0.00	**0.00**	$\mathbf{2H^+ + 2e \rightarrow H_2}$
	+0.34	$Cu^{2+} + 2e \rightarrow Cu$
	+0.80	$Ag^+ + e \rightarrow Ag$
+1.00		
	+1.69	$Au^+ + e \rightarrow Au$

There is a simple way of memorising this series, which must be known (it is not necessary to remember the numerical values, these will be given to you in an examination; what you must be familiar with is the relative positions of the elements).

'Little Potty Sammy Met A Mad Zebra In Lovely Honolulu Causing Strange Gazes'!

i.e.				
	Little	Lithium	Li	-3.09 V
	Potty	Potassium	K	
	Sammy	Sodium	Na	
	Met	Magnesium	Mg	
	A	Aluminium	Al	
	Mad	Manganese	Mn	
	Zebra	Zinc	Zn	
	In	Iron	Fe	
	Lovely	Lead	Pb	
	Honolulu	Hydrogen	H	0.00 V
	Causing	Copper	Cu	
	Strange	Silver	Ag	
	Gazes	Gold	Au	$+1.69$ V

The Electrochemical Series can be summarised as follows.

(a) Elements with large positive reduction potentials, $E°$, are easy to reduce and are good oxidising agents, *e.g.*

$F_2 + 2e \rightarrow 2F^- \quad E° = 2.87$ V

(b) Elements with large negative reduction potentials, $E°$, are difficult to reduce themselves, but are good reducing agents, *e.g.*

$Na^+ + e \rightarrow Na \quad E° = -2.71$ V

(c) Species with low $E°$ values reduce species with high $E°$ values and species with high $E°$ values oxidise species with low $E°$ values, *i.e.* low reduces high and **h**igh **o**xidises **l**ow (**HOL**).

$Zn_{(s)} + Cu^{2+}{}_{(aq)} \rightarrow Cu_{(s)} + Zn^{2+}{}_{(aq)}$.
$E°(Zn^{2+}, Zn) = -0.76$ V; $E°(Cu^{2+}, Cu) = +0.34$ V.

Reduction potentials vary in a complicated way throughout the periodic table. Generally however, the most negative are found on the left side of the table and the most positive are found on the right side. The Electrochemical Series will be discussed in greater detail in Chapter 7, with respect to electrolytic cells. In such cells, the position of an element in the activity series will determine the appropriate electrode half-cell reaction.

Thermodynamics and the Determination of *K*, the Equilibrium Constant for a Reaction

In Chapter 3 of this text, the idea of a spontaneous reaction was related to ΔG, the Gibbs free energy change of a reaction.

In particular,

ΔG −ve spontaneous reaction
ΔG +ve non-spontaneous reaction
$\Delta G = 0$ reaction at equilibrium

where ΔG was defined as:

$$\Delta G = \Delta H - T\Delta S$$

ΔH = change in the enthalpy (measured in J mol^{-1}), T = temperature (measured in K) and ΔS = change in the entropy (disorder) (measured in J K^{-1} mol^{-1}).

From physics, there is an expression for the voltage or potential, V: $V = w/Q \Rightarrow w = VQ$, where w is the work done (*i.e.* the energy, measured in joules, J) and Q is the charge (measured in coulombs, C).

In electrochemistry, one or multiple numbers of moles of reactants are considered, and specifically for charge:

The Faraday constant is 96 500 C mol^{-1}
i.e. the charge of 1 mole of electrons = F = 96 500 C mol^{-1}
$\Rightarrow$ electrical energy = potential × charge (*i.e.* $w = VQ$)

$\Rightarrow \Delta G = E_{rev}(-\nu F)$, where ΔG, the change in Gibbs free energy is, in fact, the change in electrical potential energy. This is normally written as $\Delta G = -\nu FE_{rev}$ where rev indicates reversibility.

This leads to a very useful expression, which connects thermodynamics calculations to electrochemistry, and moreover to the determination of K, the equilibrium constant:

$$\Delta G = |-\nu FE|$$

where ν = the number of electrons participating in the reaction, as defined by the equation describing the half-cell reaction (*e.g.* $Fe^{2+}_{(aq)} + 2e \rightarrow Fe^0_{(s)}$; $\nu = 2$); F = the Faraday constant = 96 500 C mol^{-1}; E = the electrode potential.

At standard state conditions, *i.e.* 25 °C and 1 bar pressure, the expression becomes:

$$\boxed{\Delta G^\circ = -\nu FE^\circ_{\text{cell}}}$$

where $E^\circ_{\text{cell}} = E^\circ_{\text{RHE}} - E^\circ_{\text{LHE}}$.

From this, one of the most important equations in electrochemistry, the ***Nernst equation***, can be derived:
$\Delta G = \Delta G^\circ + RT \ln K$, where R = Universal Gas Constant = 8.314 $\text{J K}^{-1}\,\text{mol}^{-1}$.

$$\text{But, } \Delta G^\circ = -\nu FE^\circ_{\text{cell}}$$
$$\Rightarrow -\nu FE = -\nu FE^\circ_{\text{cell}} + RT \ln K$$

Dividing across by $-\nu F$:

$$\boxed{E = E^\circ_{\text{cell}} - (RT/\nu F) \ln K, \textit{ the Nernst equation}}$$

For the general reaction: $\nu_A\text{A} + \nu_B\text{B} \rightarrow \nu_C\text{C} + \nu_D\text{D}$, where ν_A, ν_B, ν_C and ν_D represent the stoichiometry factors, K, the equilibrium constant can be written as follows:

$$K = \frac{[C]^{\nu_C}[D]^{\nu_D}}{[A]^{\nu_A}[B]^{\nu_B}}$$

i.e. of the form 'products/reactants'
e.g. for the reaction $\text{Fe}^{\text{III}}_{(\text{aq})} + \text{SCN}^-_{(\text{aq})} \rightleftharpoons \text{Fe}^{\text{II}}(\text{NCS})^{2+}_{(\text{aq})}$,
$K = [\text{Fe}^{\text{II}}(\text{NCS})^{2+}_{(\text{aq})}]/\{[\text{Fe}^{\text{III}}_{(\text{aq})}][\text{SCN}^-_{(\text{aq})}]\}$

As mentioned in Chapter 4, the activity, a, of a solid is unity, *i.e.* on examination of the balanced chemical equation, any substance with an (s) subscript implies the substance is in the solid state and therefore has unit activity.

At equilibrium: $\Delta G = 0 \quad \Rightarrow -\nu FE = 0 \quad \Rightarrow E = 0$
$\Rightarrow$ For the Nernst equation: $E = E^\circ_{\text{cell}} - (RT/\nu F)\ln K$,
i.e. $0 = E^\circ_{\text{cell}} - (RT/\nu F)\ln K$
$\Rightarrow (RT/\nu F)\ln K = E^\circ_{\text{cell}} \Rightarrow \ln K = (\nu FE^\circ_{\text{cell}}/RT)$,

from which K, the equilibrium constant can be determined.

In conclusion, three equations should be remembered:

(a) $E°_{cell} = E°_{RHE} - E°_{LHE}$
(b) $\Delta G° = -\nu FE°_{cell}$
(c) $E = E°_{cell} - (RT/\nu F)\ln K$ (the Nernst equation)
Do not forget that at equilibrium, $E = 0$, since $\Delta G = 0$,
i.e. $\ln K = (\nu FE°_{cell})/(RT)$

WORKING METHOD FOR GALVANIC CELL PROBLEMS

The following is an outline of the stepwise procedure on how to approach a problem on a galvanic cell:

1. Read the question very carefully. Remember, if you see the words 'electrolysis' or 'electrolysed', this refers to an electrolytic cell and *not* a galvanic cell. The working method which follows is only applicable to galvanic cells.
2. From the one-line representation of the cell, the balanced chemical equation *or* the two standard electrode potentials, identify the two *types of electrodes* involved. For example:

   (a) Metal–metal-ion electrode, *e.g.* $Fe_{(s)}|Fe^{2+}{}_{(aq)}$;
   (b) Metal ion in two different valence states, *e.g.* $Pt_{(s)}|Fe^{3+}{}_{(aq)}, Fe^{2+}{}_{(aq)}$;
   (c) Gas-ion electrode, *e.g.* $Pt|H_{2(g)}|H^{+}{}_{(aq)}$;
   (d) Metal – insoluble salt anion electrode, *e.g.* $Ag_{(s)}|AgCl_{(s)}|Cl^{-}{}_{(aq)}$.

3. You should now re-examine the problem. This is probably the most important or most critical step in the working method.
   (a) If the $E°$ values *alone* are given (*i.e.* if a balanced chemical equation is *not given* in the question), then the more positive $E°$ value will indicate the electrode acting as the cathode. Remember, $E°$ values are standard reduction potentials and reduction in any electrochemical cell (both galvanic and electrolytic) takes place at the cathode ('CROA').
   (b) If a balanced chemical equation is given in the question, you now have to determine, from the oxidation numbers, which species is *oxidised* and which species is *reduced*, remembering:
   Reduction is a decrease in the oxidation number,
   e.g. $Fe^{3+}{}_{(aq)} + e \rightarrow Fe^{2+}{}_{(aq)}$
   Oxidation is an increase in the oxidation number,
   e.g. $Ce^{3+}{}_{(aq)} \rightarrow Ce^{4+}{}_{(aq)} + e$
4. Once you have determined which electrode is acting as the

cathode and which electrode is acting as the anode in the cell, write down the *one-line representation* of the cell, and follow convention by keeping the cathode as the right-hand electrode:

LHE	‖	RHE
Anode		Cathode

Single vertical lines, (|) should be used to represent phase boundaries or junctions and a double line (‖) should be used in the centre to separate the two half-cells, which can be a salt bridge (*e.g.* KNO_3).

5. Write down the two electrode *half-reactions*, remembering:

   (a) 'CROA': **c**athode–**r**eduction, **a**node–**o**xidation
   (b) 'OILRIG': **o**xidation **i**s **l**oss of electrons, **r**eduction **i**s **g**ain of electrons
   (c) Reduction is a decrease in the oxidation number, **o**xidation is an **i**ncrease in the oxidation number (recall the two vowels!)

6. It may now be necessary to *balance* these two half-reactions, such that the number of electrons lost, or the number of electrons gained, is identical for both. This will then yield ν, the number of electrons in the Nernst equation: $E = E°_{cell} - (RT/\nu F)\ln K$.
   The two balanced electrode half-reactions should now be added, to give the net cell reaction. The electrons should cancel each other on both sides of the net cell reaction.
7. Draw the cell, and indicate clearly:

   (a) The cathode (RHE) and the anode (LHE).
   (b) The salt bridge, *e.g.* KNO_3 (if present).
   (c) The direction of electron flow in the wire or solution (remember electrons move from the anode to the cathode, just remember the vowels again!).
   (d) The spontaneous direction of current, I (simply the opposite direction to the movement of electrons in the wire).
   (e) The ion flow: $-$ve ions from the salt bridge migrate to the anode and $+$ve ions from the salt bridge migrate to the cathode.

   Figure 6.8 shows a typical galvanic cell diagram.
8. Determine $E°_{cell}$ from the equation $E°_{cell} = E°_{RHE} - E°_{LHE}$,

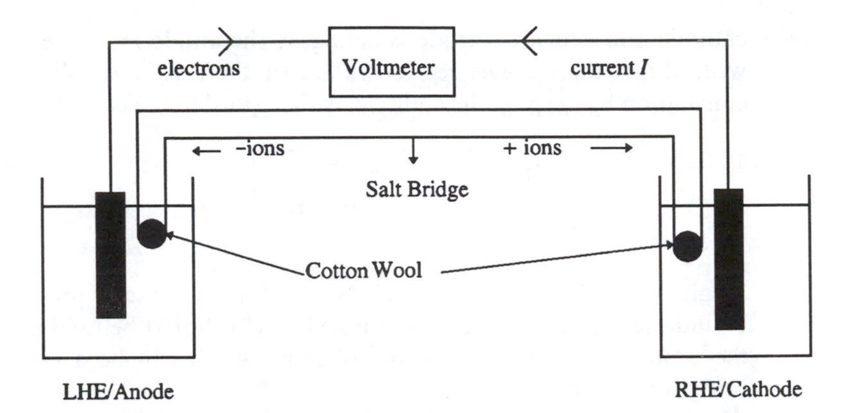

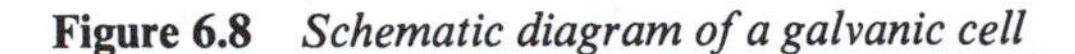
Figure 6.8 *Schematic diagram of a galvanic cell*

where the RHE is the cathode. Note that if the LHE transpires to be more positive than the RHE, this signifies a non-spontaneous cell, as normally $E°_{RHE}$ will be greater (*i.e.* more +ve) than $E°_{LHE}$ (more −ve) for a spontaneous galvanic cell.

9. Apply the Nernst equation:
   $E = E°_{cell} - (RT/\nu F)\ln K$, where: R = Universal Gas Constant = 8.314 J K^{-1} mol^{-1} (given in question); T = temperature in K (not °C) (standard state conditions imply 298 K, *i.e.* 25 °C and 1 bar pressure); F = 96 500 C mol^{-1} (given in question); $E°_{cell} = E°_{RHE} - E°_{LHE}$ in V, obtained from step 8; K = equilibrium constant; ν = defined and determined previously in step 6. Remember also that at equilibrium, $E = 0$.
10. Determine E, K or whatever unknown parameter is required, and do not forget to *state the units!*
11. Answer any riders to the question, *e.g.* determination of the solubility product, K_{sp}, *etc.*

Examples 1–4 now apply this working method.

Example No. 1: Draw the cell represented by:
$Pt_{(s)}|H_{2(g)}|H^+_{(aq)}||Cl^-_{(aq)}|AgCl_{(s)}|Ag_{(s)}$
where $E°$ $(Ag_{(s)}|AgCl_{(s)}|Cl^-_{(aq)})$ = +0.222 V. If the cell was short-circuited, indicate clearly on the diagram the cathode, the anode, the direction of spontaneous current, the electron flow and the ion flow. Write down the two respective half-reactions and the net cell reaction. Determine the equilibrium constant of the reaction at 298 K, given that R = 8.314 J K^{-1} mol^{-1} and F = 96 500 C mol^{-1}.

Solution:

1. There is no mention of 'electrolysis' or 'electrolysed'; therefore the cell in question is a galvanic cell.
2. $Pt_{(s)}|H_{2(g)}|H^+{}_{(aq)} \parallel Cl^-{}_{(aq)}|AgCl_{(s)}|Ag_{(s)}$
   (a) LHE is a gas–ion electrode, namely the standard hydrogen electrode (SHE).
   (b) RHE is a metal–insoluble salt anion electrode.
3. No balanced chemical equation is given. Hence, the $E°$ values can be used to determine which electrode is acting as the cathode: $E°(Ag_{(s)}|AgCl_{(s)}|Cl^-{}_{(aq)}) = +0.222$ V; $E°\ (H^+{}_{(aq)}|H_{2(g)}) =$ 0.000 V (the SHE; value not given but should be known).
   Therefore $Ag_{(s)}|AgCl_{(s)}|Cl^-{}_{(aq)}$ acts as the cathode as this is the more positive $E°$ value, and the SHE is the anode.
4. One-line representation of the cell:
   $Pt_{(s)}|H_{2(g)}|H^+{}_{(aq)}||Cl^-{}_{(aq)}|AgCl_{(s)}|Ag_{(s)}$
5. Anode reaction (SHE): $H_{2(g)}$ (0); $H^+{}_{(aq)}$ (I).
   Hence, 0 →1 (oxidation), *i.e.* $H_{2(g)} \rightarrow H^+{}_{(aq)} + e$ oxidation
   Therefore, $H_{2(g)} \rightarrow 2H^+{}_{(aq)} + 2e$ or $\frac{1}{2}H_{2(g)} \rightarrow H^+{}_{(aq)} + e$
   Cathode reaction: $Ag_{(s)}$ (0); $Cl^-{}_{(aq)}$ (−I); $AgCl_{(s)}$: Ag (I); Cl (−I).
   There is no change in the oxidation state of the chlorine (−I → −I), but there is a change in the oxidation state of the silver (I → 0).

Therefore: I → 0 reduction

$AgCl_{(s)} \rightarrow Ag_{(s)}$

$\Rightarrow \quad AgCl_{(s)} + e \rightarrow Ag_{(s)} + Cl^-{}_{(aq)}$

I −I → 0 −I

Anode reaction: $\frac{1}{2}H_{2(g)} \rightarrow H^+{}_{(aq)} + e$

Cathode reaction: $AgCl_{(s)} + e \rightarrow Ag_{(s)} + Cl^-{}_{(aq)}$

6 Anode reaction: $\frac{1}{2}H_{2(g)} \rightarrow H^+{}_{(aq)} + e$

Cathode reaction: $AgCl_{(s)} + e \rightarrow Ag_{(s)} + Cl^-{}_{(aq)}$

Cell reaction: $AgCl_{(s)} + \frac{1}{2}H_{2(g)} \rightleftharpoons H^+{}_{(aq)} + Ag_{(s)} + Cl^-{}_{(aq)}$

7. Diagram of the cell (Figure 6.9)
8. $E°_{cell} = E°_{RHE} - E°_{LHE} = (+0.222) - (0.000)\ V = +0.222$ V.

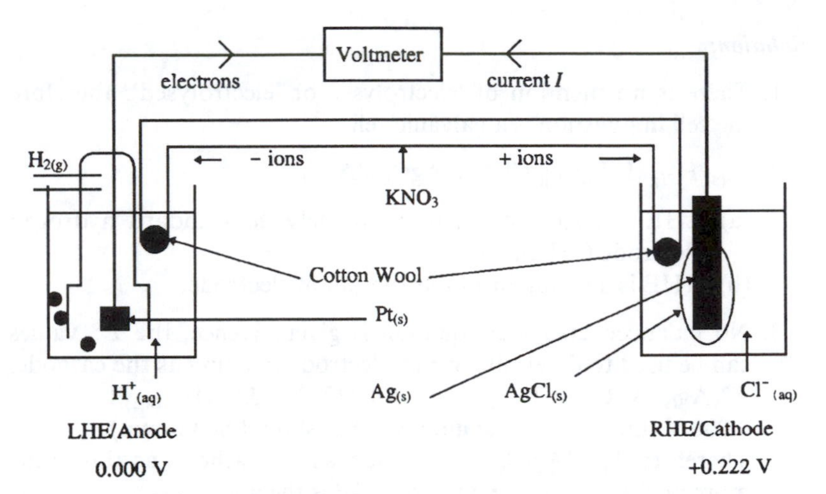

Figure 6.9 *Cell for Example No. 1*

9. Nernst equation: $E = E°_{cell} - (RT/\nu F)\ln K$.

$$AgCl_{(s)} + \tfrac{1}{2}H_{2(g)} \rightleftharpoons H^+_{(aq)} + Ag_{(s)} + Cl^-_{(aq)}$$

$K = \{[H^+_{(aq)}][Cl^-_{(aq)}]\}/[H_{2(g)}]^{1/2}$, since the activity, a, of both $AgCl_{(s)}$ and $Ag_{(s)}$ is unity (both are solid). At equilibrium:
$\Delta G = 0 \Rightarrow -\nu FE = 0 \Rightarrow E = 0$
$\Rightarrow$ For the Nernst equation:
$E = E°_{cell} - (RT/\nu F)\ln K$: $0 = E°_{cell} - (RT/\nu F)\ln K$
$\Rightarrow \ln K = (\nu F/RT)E°_{cell}$
$= [(1 \times 96\,500) / (8.314 \times 298)] \times 0.222$
$= 8.6467719$
$\Rightarrow K = \exp 8.6467719 = 5.6917 \times 10^3$.

Answer: $K = 5.6917 \times 10^3$

Example No. 2: Give a fully-labelled diagram of the galvanic cell represented by:
$Pb_{(s)}|PbSO_{4(s)}|\ SO_4{}^{2-}{}_{(aq)}$ (0.219 M)$||Sn^{4+}{}_{(aq)}$ (0.349 M), $Sn^{2+}{}_{(aq)}$ (0.248 M),$|Pt_{(s)}$. If the cell was short-circuited, indicate clearly on the diagram the cathode, the anode, the direction of spontaneous current, the electron flow and the ion flow. Write down the two respective half-reactions and the net cell reaction, where $E°$ ($Pb_{(s)}|PbSO_{4(s)}|SO_4{}^{2-}{}_{(aq)}$) $= -0.356$ V and $E°$ ($Sn^{4+}{}_{(aq)}$, $Sn^{2+}{}_{(aq)}$) $= +0.154$ V. Given that $R = 8.314$ J K^{-1} mol^{-1} and $F = 96\,500$ C mol^{-1}, determine E, the theoretical cell potential. Suggest also another standard method of analysis for the sulfate oxyanion, $SO_4{}^{2-}$, other than $PbSO_4\downarrow$.

Solution:

1. There is no mention of 'electrolysis' or 'electrolysed'; therefore the cell in question is a galvanic cell.
2. $Pb_{(s)}|PbSO_{4(s)}|SO_4{}^{2-}{}_{(aq)}||Sn^{4+}{}_{(aq)}$, $Sn^{2+}{}_{(aq)}|Pt_{(s)}$. This time, identifying the two types of electrodes is more difficult. Remember however, all you need do is determine which of the four types of electrode is involved in each case.

   (a) $Pb_{(s)}|PbSO_{4(s)}|SO_4{}^{2-}{}_{(aq)}$ is a metal–insoluble salt anion electrode;
   (b) $Sn^{4+}{}_{(aq)}$, $Sn^{2+}{}_{(aq)}|Pt_{(s)}$ is a metal ion in two different valence states (note the inert metal, Pt).

3. No balanced chemical equation is given, therefore the $E°$ values can be used to determine directly which electrode is acting as the cathode:
   $E°\ (Pb_{(s)}|PbSO_{4(s)}|SO_4{}^{2-}{}_{(aq)}) = -0.356\ V$;
   $E°\ (Sn^{4+}{}_{(aq)}, Sn^{2+}{}_{(aq)}) = +0.154\ V$.
   The latter has the more positive value ⇒ cathode.
4. One-line representation of the cell:
   $Pb_{(s)}|PbSO_{4(s)}|SO_4{}^{2-}{}_{(aq)}\ ||Sn^{4+}{}_{(aq)}, Sn^{2+}{}_{(aq)}|Pt_{(s)}$ (Figure 6.10).
5. Cathode reaction: ('CROA') $Sn^{4+}{}_{(aq)}$ (IV); $Sn^{2+}{}_{(aq)}$ (II). Hence, 4 → 2 (reduction), *i.e.* $Sn^{4+}{}_{(aq)} + 2e \rightarrow Sn^{2+}{}_{(aq)}$;
   Anode reaction: $Pb_{(s)}$ (0); $PbSO_{4(s)}$ has Pb (II) and $SO_4{}^{2-}$ (−II); also $SO_4{}^{2-}$ (−II) present. There is no change in the oxidation state of the sulfur or oxygen, but the oxidation state of the lead does change (0 → II), *i.e.* $Pb^0{}_{(s)} \rightarrow Pb^{2+}{}_{(aq)} + 2e$.
   Therefore, $Pb^0{}_{(s)} + SO_4{}^{2-}{}_{(aq)} \rightarrow PbSO_{4(s)} + 2e$
6. Cathode reaction: $Sn^{4+}{}_{(aq)} + 2e \rightarrow Sn^{2+}{}_{(aq)}$
   Anode reaction: $Pb_{(s)} + SO_4{}^{2-}{}_{(aq)} \rightarrow PbSO_{4(s)} + 2e$

   Cell reaction:
   $Sn^{4+}{}_{(aq)} + Pb_{(s)} + SO_4{}^{2-}{}_{(aq)} \rightarrow PbSO_{4(s)} + Sn^{2+}{}_{(aq)}$
7. Diagram of the cell (Figure 6.10)
8. $E°_{cell} = E°_{RHE} - E°_{LHE} = (+0.154) - (-0.356)\ V = +0.510\ V$.
9. Nernst equation: $E = E°_{cell} - (RT/\nu F) \ln K$.

   $$Sn^{4+}{}_{(aq)} + Pb_{(s)} + SO_4{}^{2-}{}_{(aq)} \rightleftharpoons PbSO_{4(s)} + Sn^{2+}{}_{(aq)}$$

   $K = [Sn^{2+}{}_{(aq)}]/\{[Sn^{4+}{}_{(aq)}][SO_4{}^{2-}{}_{(aq)}]\}$, since the activity, a, of both $Pb_{(s)}$ and $PbSO_{4(s)}$ is unity (both are solid).
   $\Rightarrow E = 0.510 - [(8.314 \times 298)/(2 \times 96\,500)] \times \ln[0.248/(0.349 \times 0.219)] = +0.495\ V$.

Answer: $\boldsymbol{E = +0.495\ V}$

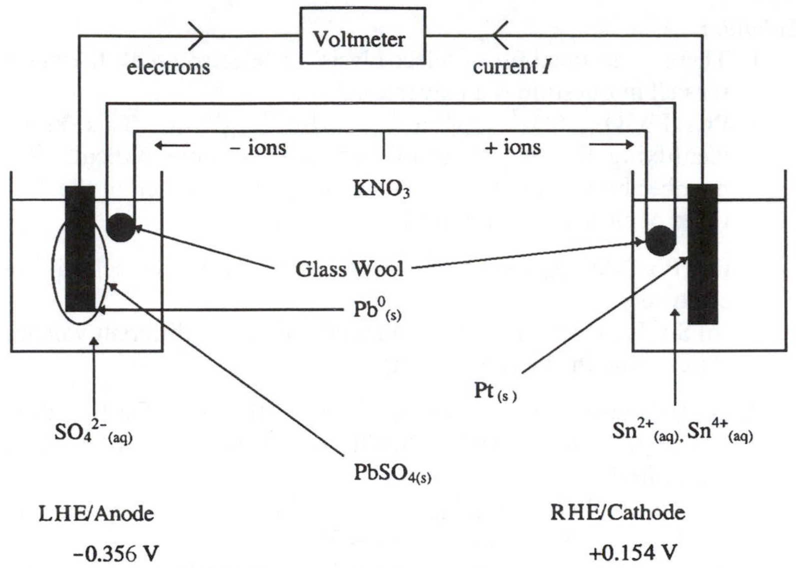

Figure 6.10 *Cell for Example No. 2*

10. Precipitation reaction: $Ba^{2+}{}_{(aq)} + SO_4{}^{2-}{}_{(aq)} \rightarrow BaSO_{4(s)}\downarrow$.

Here are some other important precipitation reactions which you are liable to meet in your chemistry course:

(a) $Ag^+ + Cl^- \rightarrow AgCl\downarrow$;
$Ag^+ + Br^- \rightarrow AgBr\downarrow$;
$Ag^+ + I^- \rightarrow AgI\downarrow$

(b) $Ca^{2+}{}_{(aq)} + SO_4{}^{2-}{}_{(aq)} \rightarrow CaSO_{4(s)}\downarrow$;
$Ca^{2+}{}_{(aq)} + C_2O_4{}^{2-}{}_{(aq)} \rightarrow CaC_2O_{4(s)}\downarrow$;
$Ca^{2+}{}_{(aq)} + CO_3{}^{2-}{}_{(aq)} \rightarrow CaCO_{3(s)}\downarrow$;

(c) $Ba^{2+}{}_{(aq)} + SO_3{}^{2-}{}_{(aq)} \rightarrow BaSO_{3(s)}\downarrow$

Example No. 3: Give a fully labelled diagram of the galvanic cell based on the following redox reaction: $Cd^0{}_{(s)} + Pb^{2+}{}_{(aq)} \rightarrow Cd^{2+}{}_{(aq)} + Pb^0{}_{(s)}$. If the cell was short-cricuited, indicate clearly on the diagram the cathode, the anode, the direction of current, the electron flow and the ion flow. Assuming the following standard electrode potential values ($Pb_{(s)}|Pb^{2+}{}_{(aq)}$, -0.126 V; $Cd_{(s)}|Cd^{2+}{}_{(aq)}$, $+0.403$ V), estimate (a) the standard potential of the cell, and (b) the EMF at 25 °C, of the following slightly modified version:

$$Pb_{(s)}|Pb^{2+}{}_{(aq)}\ (0.0008\ M)||Cd^{2+}{}_{(aq)}\ (2.32\ M)|Cd_{(s)}$$
$$(R = 8.314\ J\ K^{-1}\ mol^{-1}\ and\ F = 96\,500\ C\ mol^{-1})$$

Solution:

1. There is no mention of 'electrolysis' or 'electrolysed'; therefore the cell in question is a galvanic cell.
2. For part (a), since a balanced chemical equation is given, this determines which electrode will act as the cathode, and which electrode will act as the anode. Therefore, the oxidation state of each species must be written down:

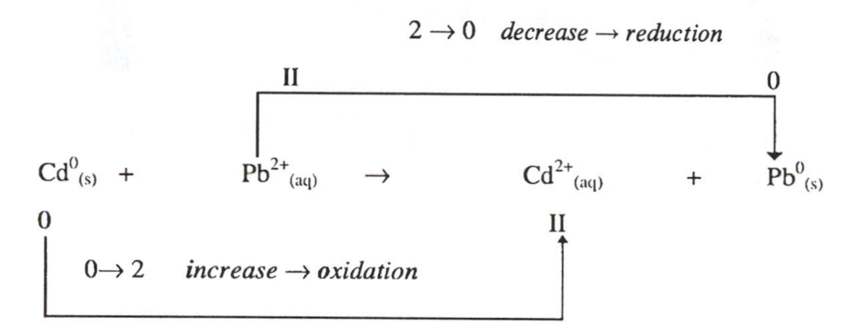

i.e. the $Pb_{(s)}|Pb^{2+}{}_{(aq)}$ electrode will act as the cathode (RHE) and the $Cd_{(s)}|Cd^{2+}{}_{(aq)}$ electrode will act as the anode (LHE), in this galvanic cell.
3. RHE is a metal–metal-ion electrode, LHE is a metal–metal ion electrode.
4. One-line representation of the cell: $Cd_{(s)}|Cd^{2+}{}_{(aq)}||Pb^{2+}{}_{(aq)}|Pb_{(s)}$
5. Cathode reaction ('CROA'): $Pb^{2+}{}_{(aq)}$ (II); $Pb_{(s)}$ (0)
   Hence, 2 → 0 (reduction) *i.e.* $Pb^{2+}{}_{(aq)} + 2e \rightarrow Pb^{0}{}_{(s)}$
   Anode reaction: Cd^{2+} (II); $Cd^{0}{}_{(s)}$ (0)
   Hence 0 → 2 (oxidation) *i.e.* $Cd^{0}{}_{(s)} \rightarrow Cd^{2+}{}_{(aq)} + 2e$
6. Cathode reaction: $Pb^{2+}{}_{(aq)} + 2e \rightarrow Pb^{0}{}_{(s)}$
   Anode reaction: $Cd^{0}{}_{(s)} \rightarrow Cd^{2+}{}_{(aq)} + 2e$

   Cell reaction: $Cd_{(s)} + Pb^{2+}{}_{(aq)} \rightarrow Cd^{2+}{}_{(aq)} + Pb_{(s)}$
7. Diagram of the cell (Figure 6.11)
8. Standard potential of the cell $E^\circ_{cell} = E^\circ_{RHE} - E^\circ_{LHE} = (-0.126) - (0.403)$ V $= -0.529$ V. But, since $\Delta G^\circ = -\nu FE^\circ = -\nu F \times -0.529)$, ΔG° is +ve, *i.e.* a non-spontaneous reaction is indicated! This is consistent with the values of the standard reduction potentials. The more positive E° value is associated with the $Cd^{0}{}_{(s)}|Cd^{2+}{}_{(aq)}$ electrode and hence in a spontaneous cell, this should act as the cathode. However, from the direction of the balanced chemical equation, it is the $Pb^{0}{}_{(s)}|Pb^{2+}{}_{(aq)}$

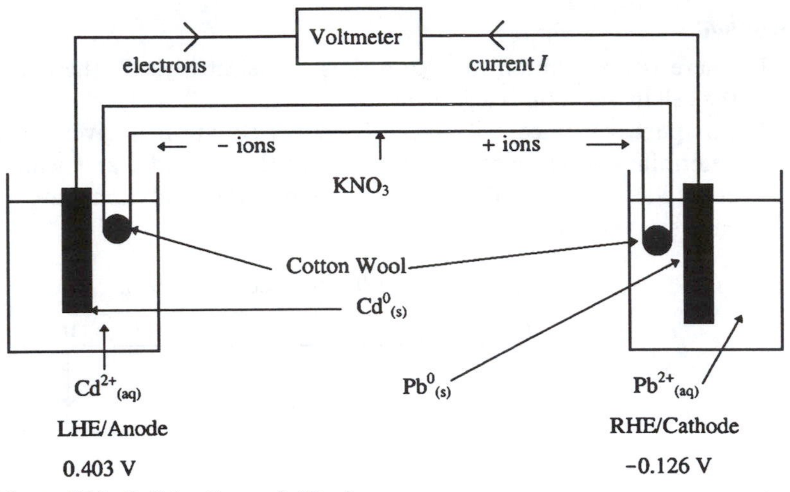

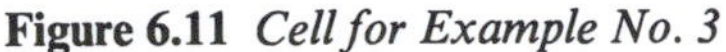

Figure 6.11 *Cell for Example No. 3*

electrode which acts as the cathode, and hence results in a non-spontaneous cell.

9. In part (b), the EMF at 25 °C of the modified cell has to be determined. In the latter case, since the $Cd^0_{(s)}|Cd^{2+}_{(aq)}$ electrode is on the right-hand side of the one-line representation of the cell, *i.e.* $Pb^0_{(s)}|Pb^{2+}_{(aq)}$ (0.0008 M)$||Cd^{2+}_{(aq)}$(2.32 M)$|Cd^0_{(s)}$, this implies that the cadmium electrode is now acting as the cathode.
   Hence: $E°_{cell} = E°_{RHE} - E°_{LHE} = (+0.403) - (-0.126)$ V = +0.529 V (spontaneous cell).
   Nernst equation: $E = E°_{cell} - (RT/\nu F)\ln K$.

   $$Pb^0_{(s)} + Cd^{2+}_{(aq)} \rightleftharpoons Pb^{2+}_{(aq)} + Cd^0_{(s)}$$

   $K = [Pb^{2+}_{(aq)}]/[Cd^{2+}_{(aq)}]$, since the activity, a, of both $Pb_{(s)}$ and $Cd_{(s)}$ is unity (since they are both solid)
   $\Rightarrow E = 0.529 - [(8.314 \times 298)/(2 \times 96\,500)] \times \ln(0.0008/2.32) = 0.631$ V.

Answer:* $E = +0.631$ *V

Example No. 4: The standard potentials of $Ag_{(s)}|Ag^+_{(aq)}$ and $Ag_{(s)}|AgCl_{(s)}|Cl^-_{(aq)}$ are +0.799 and +0.222 V respectively at 25 °C. Use this information to determine the solubility product of silver chloride. Draw the galvanic cell in question. If the cell was short-circuited, indicate clearly on the diagram the cathode, the anode, the direction of current, the electron flow and the ion flow.
($R = 8.314$ J K^{-1} mol^{-1} and $F = 96{,}500$ C mol^{-1})

Solution:

1. There is no mention of 'electrolysis' or 'electrolysed', therefore the electrochemical cell is a galvanic cell.
2. In this problem, the solubility product of silver chloride has to be determined. The solubility product equilibrium reaction first has to be written down:
   $AgCl_{(s)} \rightleftharpoons Ag^{+}{}_{(aq)} + Cl^{-}{}_{(aq)}$, where $K_{sp} = [Ag^{+}{}_{(aq)}][Cl^{-}{}_{(aq)}]$ since the activity, a, of $AgCl_{(s)}$ is unity. This is the net cell reaction that has to be obtained. From the standard electrode potentials, $Ag_{(s)}|Ag^{+}{}_{(aq)}$ has the greatest $E°$ value and therefore would act as the cathode (RHE), whereas $Ag_{(s)}|AgCl_{(s)}|Cl^{-}{}_{(aq)}$ would act as the anode (LHE) in a spontaneous galvanic cell. However, this will not generate the desired chemical equation, concerning the solubility product equilibrium of silver chloride. Put in a different way, in this question, you are given the equation in an implicit manner, and therefore, as before, you have to examine the equation to see which substance is oxidised, and which substance is reduced.
3. Cathode reaction ('CROA'): $Ag^{0}{}_{(s)}$ (0); $Cl^{-}{}_{(aq)}$ (−I); $AgCl_{(s)}$ has Ag (I) and Cl (−I). There is no change in the oxidation state of the chlorine (−I to −I), but there is a change in the oxidation state of the silver (I → 0).
   Therefore: I → 0 (reduction)
   $$AgCl_{(s)} \rightarrow Ag_{(s)}$$
   $$\Rightarrow \quad AgCl_{(s)} + e \rightarrow Ag_{(s)} + Cl^{-}{}_{(aq)}$$
   I −I → 0 −I
   Anode reaction: $Ag^{+}{}_{(s)}$ (I), $Ag^{0}{}_{(s)}$ (0)
   Hence 0 → 1 (oxidation), *i.e.* $Ag^{0} \rightarrow Ag^{+} + e$
   Hence:
   Cathode reaction: $AgCl_{(s)} + e \rightarrow Ag^{0}{}_{(s)} + Cl^{-}{}_{(aq)}$
   Anode reaction: $Ag^{0}{}_{(s)} \rightarrow Ag^{+}{}_{(aq)} + e$

   Cell reaction: $AgCl_{(s)} \rightleftharpoons Ag^{+}{}_{(aq)} + Cl^{-}{}_{(aq)}$
4. RHE is a metal-insoluble salt anion electrode, LHE is a metal–metal-ion electrode.
5. One-line representation of the cell:
   $Ag_{(s)}|Ag^{+}{}_{(aq)}||Cl^{-}{}_{(aq)}|AgCl_{(s)}|Ag_{(s)}$
6. Diagram of the cell (Figure 6.12).

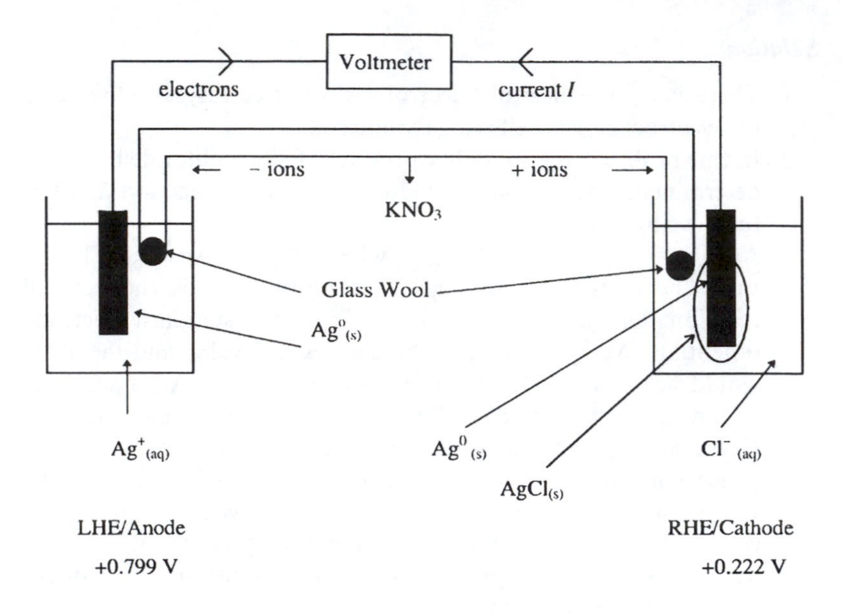

Figure 6.12 *Cell for Example No. 4*

7. Standard potential of the cell $E°_{cell} = E°_{RHE} - E°_{LHE} = (+0.222) - (0.799)\text{ V} = -0.577\text{ V}$.
8. Nernst equation: $E = E°_{cell} - (RT/\nu F)\ln K$.

   $AgCl_{(s)} \rightleftharpoons Ag^+{}_{(aq)} + Cl^-{}_{(aq)}$

   $K_{sp} = [Ag^+{}_{(aq)}][\,Cl^-{}_{(aq)}]$ since the activity, a, of AgCl is unity (since it is solid).
   At equilibrium, $E = 0 \Rightarrow \ln K_{sp} = (\nu F/RT)\,E°_{cell}$.
   Hence, $K_{sp} = \exp\,[(1 \times 96\,500 \times -0.577)/(8.314 \times 298)] = 1.7368 \times 10^{-10}$.

$$\textbf{\textit{Answer: }} K_{sp} = 1.7 \times 10^{-10}$$

At the conclusion of Chapter 7, there is a multiple-choice test, consisting of 10 questions, and three further examples for you to try. In Chapter 7, a second type of electrochemical cell is considered, the electrolytic cell.

Chapter 7

Electrochemistry II: Electrolytic Cells

ELECTROLYSIS

Galvanic cells convert chemical (potential) energy into electrical (kinetic) energy, as described in Chapter 6. ***Electrolytic cells*** convert electrical (kinetic) energy into chemical (potential) energy and therefore an electrolytic cell requires an external source of electrical energy, such as a battery, for operation. For this reason, ***electrolysis*** can be defined as the input of electrical energy from an external source, such as a battery, as direct current to force a non-spontaneous reaction to occur, *i.e.* ΔG +ve. The ***electrolyte*** is the ***solution***, which can be either an ionic or covalent compound that melts to produce ions or that dissolves to give a solution that contains ions (charged species), such as Na^+ and Cl^-. The ***electrode*** is the ***metal plate*** used to bring the electrical energy to the solution, which then brings about the chemical change in solution. ***Active electrodes*** are metals which have the same element as that contained in the solution, *i.e.* the electrolyte. A typical active electrode is copper. The copper is immersed in a solution of its own ions, $Cu^{2+}{}_{(aq)}$. ***Inactive electrodes*** consist of a metal which does not react with the solution in which it is immersed, but brings the electrical energy to the solution. Platinum and graphite are examples of inactive electrodes. ***Faraday's Laws of Electrolysis*** are the governing laws which form the background to electrolysis.

Faraday's Laws of Electrolysis

Faraday's First Law of Electrolysis states that the mass of a substance (element) deposited at the cathode ('CROA') during electrolysis is directly proportional to the quantity of charge (measured in coulombs) passing through the solution.
Faraday's Second Law of Electrolysis states that the number of moles of electrons needed to discharge one mole of an ion at an electrode is equal to the number of charges on that ion.

In summary : $m \propto Q$

$m \propto It$ (*since $Q = It$ from physics*)

$\Rightarrow m = kIt$

(*since a proportionality sign can always be replaced by '= k'*)

$\Rightarrow m = zIt$

where z is defined as the electrochemical equivalent of the element.

The ***electrochemical equivalent*** of an element is the mass of that element produced (deposited, in the case of a solid, or evolved, in the case of a gas) at the cathode when 1 coulomb of charge passes through the electrolyte solution.

Note the correct units: I = current, measured in ampères, A; t = time, measured in seconds, s; Q = charge, measured in coulombs, C; m = mass, measured in kilograms, kg (the unit of mass is the kilogram, kg, not the gram, g). Hence, since $m = zIt$, $z = m/(It)$ and so kg A^{-1} s^{-1} is the unit of the electrochemical equivalent.

The quantity of electricity can then be measured in terms of the number of moles of electrons passing through the electrolytic cell. The amount of substance undergoing the chemical change is related to the number of electrons involved in the respective half-reaction, and can be expressed in terms of the number of moles of substance or the number of reactive species, *i.e.* the number of chemical equivalents. The concept of a redox reaction and the number of reactive species is used to determine the amount of substance deposited or the volume of gas evolved during electrolysis, *e.g.*

$$Cu^{2+}{}_{(aq)} + 2e \rightarrow Cu^{0}{}_{(s)}$$

$\Rightarrow$ 2 electrons are needed to reduce $Cu^{2+}{}_{(aq)}$ to metallic copper, $Cu^{0}{}_{(s)}$.

⇒ 2 mol of electrons reduce 1 mol of copper, $Cu_{(s)}$, and 1 mol reduces 0.5 mol of copper.
⇒ the mass of 1 chemical equivalent of copper is the mass of 0.5 mol.
This leads to the definition of the faraday. The faraday is defined as the quantity of charge carried by 1 mol of electrons.

$$\mathbf{1}F = \mathbf{96\,500\ C}$$

When 1 F of electricity is passed through an electrolytic cell, 1 mole of electrons passes through the cell and 1 chemical equivalent is deposited at the cathode, *i.e.* since 'CROA' still applies in electrolytic cells, this could be expressed as 1 equivalent of substance reduced at the cathode.

⇒ 1 F = charge carried by 6.022×10^{23} (1 mol) electrons = 96 500 C

Examples:

(a) Consider the half-reaction: $Ag^{+}{}_{(aq)} + e \rightarrow Ag^{0}{}_{(s)}$
1 $F \rightarrow$ 1 mol Ag; 96 500 C → 1 mol Ag; 96 500 C → 107.868 g of silver, produced at the cathode during electrolysis, since 1 mol of Ag contains 107.868 g.

(b) Consider the half-reaction: $Mg^{2+}{}_{(aq)} + 2e \rightarrow Mg^{0}{}_{(s)}$
2 $F \rightarrow$ 1 mol Mg; 1 $F \rightarrow$ 0.5 mol Mg; 96 500 C → 0.5 mol Mg; 96 500 C → 0.5×24.305 g = 12.1525 g of magnesium produced at the cathode during electrolysis, since 1 mol of Mg contains 24.305 g.

(c) Aluminium is produced by the electrolysis of aluminium oxide, Al_2O_3 dissolved in molten cryolite, Na_3AlF_6. Determine the mass (in kg) of aluminium produced in 12 hours, in an electrolytic cell operating at 95 kA, given that the molar mass (M) of Al is 26.982 g mol^{-1} and 1 F = 96 500 C.

First, the oxidation state of Al in the oxide has to be calculated: $2x + 3(-II) = 0$; $2x = 6$; $x = III$, *i.e.* $Al^{III}{}_{(aq)}$. Then, aluminium will be deposited at the cathode (reduction), according to Faraday's Second Law of Electrolysis, *i.e.* $Al^{3+}{}_{(aq)} + 3e \rightarrow Al^{0}{}_{(s)}$
3 $F \rightarrow$ 1 mol of Al; 1 $F \rightarrow \frac{1}{3}$ mol of Al; 96 500 C → $\frac{1}{3}$ mol of Al.
But, $Q = It \Rightarrow Q = 95\,000 \times (12 \times 60 \times 60)$, since 12 hours = $12 \times 60 \times 60$ s
$\Rightarrow Q = 4.104 \times 10^{9}$ C
96 500 C → $\frac{1}{3}(26.982)$ g = 8.994 g of Al
Hence, 1 C → (8.994/96 500) g of Al
4.104×10^{9} C → $(8.994/96\,500)(4.104 \times 10^{9}) = 382501.31$ g

Answer: 382.5 kg of aluminium

Electrolytic Cells

Electrolytic cells are much easier to represent than galvanic cells, all having the standard form shown in Figure 7.1. In physics, electric current is defined as the flow of electrons from the negative pole of a battery (cathode) to the positive pole of a battery (anode) through the cell or the conventional direction of flow of current *I* from the positive pole of a battery to the negative pole through the cell. Therefore, current, by convention always goes from the positive pole of a battery to the negative pole of a battery, and electrons travel in the opposite direction (Figure 7.2).

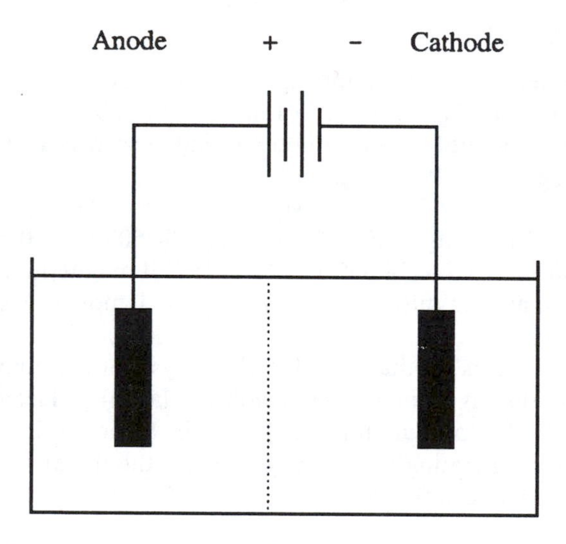

Figure 7.1 *Schematic diagram of an electrolytic cell*

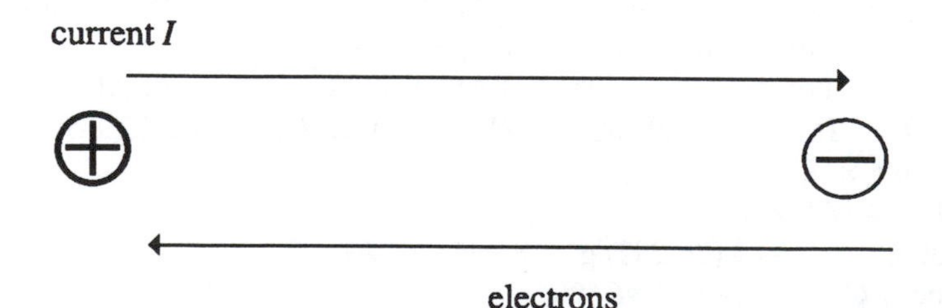

Figure 7.2 *Direction of electric current and electrons in an electrolytic cell, external to the battery*

A useful way to remember this is the mnemonic '**CNAP**', *i.e.* cathode–negative anode–positive. In electrolytic cells, the electrode connected to the negative pole of the battery is the cathode and the electrode connected to the positive pole of the battery is the anode.

> ***Summary:***
> *(a) CROA: applicable to both galvanic and electrolytic cells.*
> *(b) CNAP: applicable to electrolytic cells only.*

Figure 7.3 summarises the standard electrolytic cell.

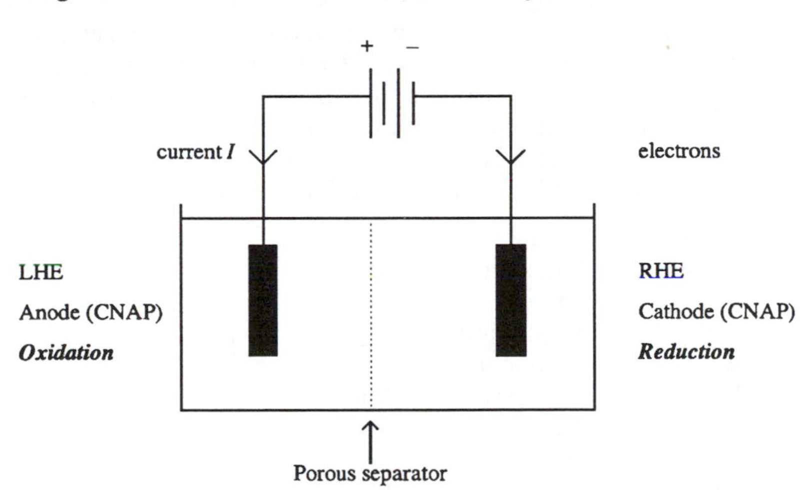

Figure 7.3 *Schematic diagram of an electrolytic cell*

The principle of electrolysis can be used to deposit metals from aqueous solutions. The next two sections describe a working method to determine the half-reactions involved in each type of the four different types of electrolytic cell.

Working Method for Electrolysis Type Problems

1. Read the question carefully. Ensure that the electrochemical cell concerned is definitely an electrolytic cell, and not a galvanic cell.
2. Is the substance to be electrolysed molten or aqueous?
3. Identify all species present. In the case of an aqueous solution, water is involved, *e.g.* in the electrolysis of a solution of sodium

chloride, $NaCl_{(aq)}$, not only are the Na^+ and the Cl^- ions present, but also H_2O, *i.e.* $H_2O \rightleftharpoons H^+ + OH^-$.

4. Determine which substances/species are present at the cathode and which substances/species are present at the anode ('CNAP'):

   Cathode is −ve, attracts +ve species
   Anode is +ve, attracts −ve species

   This procedure is modified further if water is present.

5. Examine the species at both electrodes carefully. If the electrolysis is that of a ***molten*** substance, this step can be ignored and you can go directly to step 6. If, however, the electrolysis is that of an ***aqueous*** solution, there will be a choice of two reactions at both the cathode and the anode. To determine which reduction half-reaction and which oxidation half-reaction actually occur, the following rules of thumb should be applied:

   I. Cathode reaction (reduction: $M^{n+} + ne \rightarrow M^0$):
   Write down the Electrochemical Series in detail:

Little	Lithium	Li	
Potty	Potassium	K	
Sammy	Sodium	Na	
Met	Magnesium	Mg	−ve
A	Aluminium	Al	
Mad	Manganese	Mn	
Zebra	Zinc	Zn	
In	Iron	Fe	
Lovely	Lead	Pb	
Honolulu	**Hydrogen**	**H**	**0.00 V**
Causing	Copper	Cu	
Strange	Silver	Ag	+ve
Gazes	Gold	Au	

   (a) If the element is ***above zinc*** in the Electrochemical Series, then ***$H_{2(g)}$ will be discharged*** at the cathode, according to the following reduction half-reaction:

$$2H_2O_{(aq)} + 2e \rightarrow H_{2(g)} + 2OH^-_{(aq)}$$

$2H_2O_{(aq)}$	+	2e	→	$H_{2(g)}$	+	$2OH^-_{(aq)}$
I −II				0		−II I

$1 \rightarrow 0$ (*reduction*)

(b) If the species is ***below zinc***, the positive M^{z+} cation will be reduced to the ***metal***:

$$e.g.\ Ag^{+}_{(aq)} + e \rightarrow Ag^{0}_{(s)}$$

(c) If the electrolysis involves an acid (*i.e.* a proton donor), write down the appropriate ionisation reaction for the acid with water. This will then generate H_3O^+ ions at the cathode:

$$e.g.\ H_2O + H_2SO_4 \rightarrow H_3O^+ + HSO_4^-$$
$$H_2O + HSO_4^- \rightarrow H_3O^+ + SO_4^{2-}$$

II. Anode reaction (oxidation):
Examine the position of the anion in the following series:

$$I^- > OH^- > Cl^- > NO_3^- > SO_4^{2-}$$

$\Vert$ (under Cl^-)

$\longleftarrow$

Ease of Oxidation

Anions on the left-hand side of the above series such as iodide will be easily oxidised, whereas those on the right-hand side, such as the sulfate oxyanion, SO_4^{2-}, will not be easily oxidised. In sulfate, S is in a (VI) oxidation state, having the stable [Ne]s^0p^0 inert gas core configuration (*cf.* S(0): [Ne]$3s^23p^4$). For this reason, the oxidation of water, H_2O, will occur instead, with oxygen gas evolved at the anode:

$$2H_2O_{(aq)} \rightarrow O_{2(g)} + 4e + 4H^+_{(aq)}$$

$2H_2O_{(aq)}$	$\rightarrow$	$O_{2(g)}$	+	4e	+	$4H^+_{(aq)}$
I −II		0				I

$-2 \rightarrow 0$ (*increase in oxidation number* $\Rightarrow$ *oxidation*)

If the anion is chloride, this 'sits on the fence' within the series, and which oxidation occurs is concentration dependent, *i.e.* if the solution is concentrated, chloride will be oxidised to chlorine gas, according to the reaction: $2Cl^-_{(aq)} \rightarrow Cl_{2(g)} + 2e$. If, however, the solution is dilute, water will be more easily oxidised, and the half-reaction above will take place. All this relates to the $E°$ values. In the case of chloride, $E° = -1.36$ V, and in the case of water the corresponding value is -1.23 V. As stated previously, it is not necessary to memorise these values. All that is required is a knowledge of the relative position of each species within the Electrochemical Series, as shown in Table 6.2 of Chapter 6.

6. Having established the exact reduced species and the exact oxidised species at the electrodes, write down the cathode and anode half-reactions, remembering (as for galvanic cells) that reduction takes place at the cathode, and that oxidation takes place at the anode, *i.e.* 'CROA' and 'OILRIG'.
7. Balance the two half-reactions, so that the number of electrons transferred is the same for both, and from these reactions, determine the net cell reaction
8. Draw the electrolytic cell (Figure 7.3).
   Label clearly: (a) the anode (LHE) and the cathode (RHE); (b) the direction of current, I; (c) The direction of electrons; (d) the direction of ion flow, as derived from the half-cell reactions in step 6.
9. Re-read the question, to see exactly what you are asked to determine.
10. Apply Faraday's Second Law of Electrolysis:

$$\begin{aligned} & M^{n+} + ne \rightarrow M^0 \\ \Rightarrow\ & nF \rightarrow 1 \text{ mol of M} \\ & 1\,F \rightarrow 1/n \text{ mol of M} \\ & 96\,500 \text{ C} \rightarrow 1/n \text{ moles of M} \end{aligned}$$

Also:

$$\boxed{Q = It}$$

where Q = charge (measured in coulombs, C), I = current (measured in ampères, A); t = time (measured in seconds, s).

11. Answer any riders to the question:

   (a) Standard state conditions: most stable state at 25 °C (298 K) and 1 bar pressure.
   (b) 1 mol of an ideal gas at 25 °C (298 K) and 1 bar pressure occupies 24.8 dm^3.
   (c)

$$\frac{p_1 V_1}{T_1} = \frac{p_2 V_2}{T_2}$$

*Remembered by '**P**eas and **V**egetables go on the **T**able!'*

   (d) $pH = -\log_{10}[H_3O^+]$; $pOH = -\log_{10}[OH^-]$; $pH + pOH = 14$

TYPES OF ELECTROLYTIC CELLS

Electrolysis of Molten NaCl

1. Determine whether or not the electrolysis involves (a) a molten (melted) or (b) an aqueous substance. In this example, the electrolyte is molten sodium chloride.
2. Identify all species present. If the electrolysis involves a molten substance, this step is easy, since the species present are simply the ions of the substance, *i.e.* water is not involved. Molten sodium chloride contains equal numbers of sodium Na^+ cations and chloride Cl^- anions, respectively *i.e.*
$$NaCl_{(l)} \rightarrow Na^+_{(l)} + Cl^-_{(l)}$$
3. Having identified all species present, determine which species are attracted to the cathode, and which species are attracted to the anode, remembering that the cathode is negative and the anode is positive ('CNAP'), and also that like charges repel each other and unlike charges attract one another.
$$\text{Cathode} -\text{ve} \quad Na^+ \qquad \text{Anode} +\text{ve} \quad Cl^-$$
4. Determine the two respective half-reactions, recalling that reduction takes place at the cathode and oxidation takes place at the anode ('CROA').

Cathode reaction: $Na^+_{(l)} + e \rightarrow Na^0_{(l)}$
Anode reaction: $Cl^-_{(l)} \rightarrow \frac{1}{2} Cl_{2(g)} + e$

Cell reaction: $Na^+_{(l)} + Cl^-_{(l)} \rightarrow Na^0_{(l)} + \frac{1}{2} Cl_{2(g)}$

In the electrolysis of molten sodium chloride, a pool of sodium metal deposits at the cathode, and bubbles of chlorine gas form at the anode.

5. Draw the cell (Figure 7.4). The porous separator permits the diffusion of ions from one side of the cell to the other, and prevents the sodium produced at the cathode reacting with the chlorine produced at the anode. The experimental set-up above is used commercially in the ***Downs cell*** for the electrolysis of molten sodium chloride.

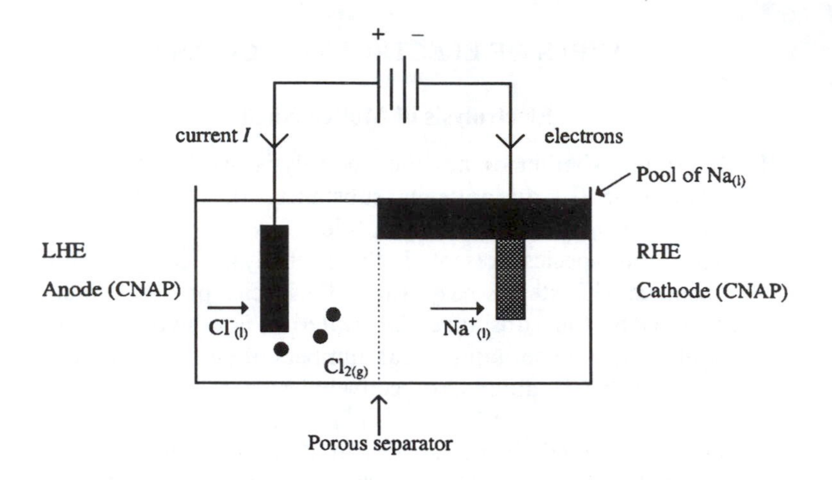

Figure 7.4 *The Downs cell for the electrolysis of molten sodium chloride*

Electrolysis of Concentrated Aqueous NaCl

1. Determine whether or not the electrolysis involves (a) a molten salt or (b) an aqueous solution of a salt. In this example, the electrolysis involves aqueous sodium chloride. Therefore, there is now an additional factor to be considered, water!
2. Identify all species present: $Na^+{}_{(aq)}$, $Cl^-{}_{(aq)}$ and H_2O.
3. Having identified all the species present, determine which species accumulate at the cathode, and which species accumulate at the anode:

   Cathode −ve Na^+, H_2O Anode +ve Cl^-, H_2O
4. The question now to be asked is which half-reaction occcurs at the cathode and which half-reaction occurs at the anode, because there are two possible half-reactions at each electrode. Before determining which half-reaction occurs, write down all possible half-reactions. In this example, this step is carried out, but a very quick way to determine which half-reaction takes precedence is subsequently provided, using the Electrochemical Series and a convenient rule of thumb.

   At the cathode: −ve electrode ('CNAP')/reduction takes place here ('CROA'/'OILRIG').
   There are two possible half-reactions:

   (a) $Na^+{}_{(aq)} + e \rightarrow Na_{(l)}$ $E^\circ = -2.714$ V.

   I 0

(b) $2H_2O_{(aq)} + 2e \rightarrow H_{2(g)} + 2OH^-_{(aq)}$ $E° = -0.08$ V.
I −II 0 −II I

1 to 0 (*reduction*)

In order to determine which half-reaction will occur, the $E°$ values need to be considered. Since the standard electrode potential for sodium is −2.714 V, and that for the water is −0.08 V, this means that water, having the relatively greater $E°$ value, will be more easily reduced at the cathode than $Na^+_{(aq)}$, and therefore the half-reaction (b) will dominate.

At the anode: +ve electrode ('CNAP')/oxidation takes place here ('CROA'/'OILRIG').
There are two possible half-reactions:

(a) $Cl^-_{(aq)} \rightarrow \frac{1}{2}Cl_{2(g)} + e$ $E° = -1.36$ V.
−I 0

(b) $2H_2O_{(aq)} \rightarrow O_{2(g)} + 4H^+_{(aq)} + 4e$ $E° = -1.23$ V.
I −II 0 I

−2 to 0 (*oxidation*)

This time, since $E°$ for the chloride is −1.36 V and $E°$ for the water is −1.23 V, the values are so close that it will be concentration dependent as to which half-reaction will actually occur, *i.e.* if the solution is concentrated sodium chloride, the concentration of chloride ion will be large, and chlorine gas will be evolved at the anode, but if the solution is very dilute, then oxygen gas will be evolved at the anode. In this example, let us assume that the sodium chloride is very concentrated:

Cathode reaction: $2H_2O_{(aq)} + 2e \rightarrow H_{2(g)} + 2OH^-_{(aq)}$
Anode reaction: $2Cl^-_{(aq)} \rightarrow Cl_{2(g)} + 2e$

Cell reaction: $2H_2O_{(aq)} + 2Cl^-_{(aq)} \rightarrow H_{2(g)} + Cl_{2(g)} + 2OH^-_{(aq)}$
In the electrolysis of concentrated aqueous sodium chloride, hydrogen gas, $H_{2(g)}$ is evolved at the cathode along with hydroxide anion, OH^- (link to pH), and chlorine gas, $Cl_{2(g)}$, is evolved at the anode.

5. Draw the cell (Figure 7.5). In this cell, the sodium cation, $Na^+_{(aq)}$ is termed a ***spectator ion***, as it does not participate in the reduction half-reaction.

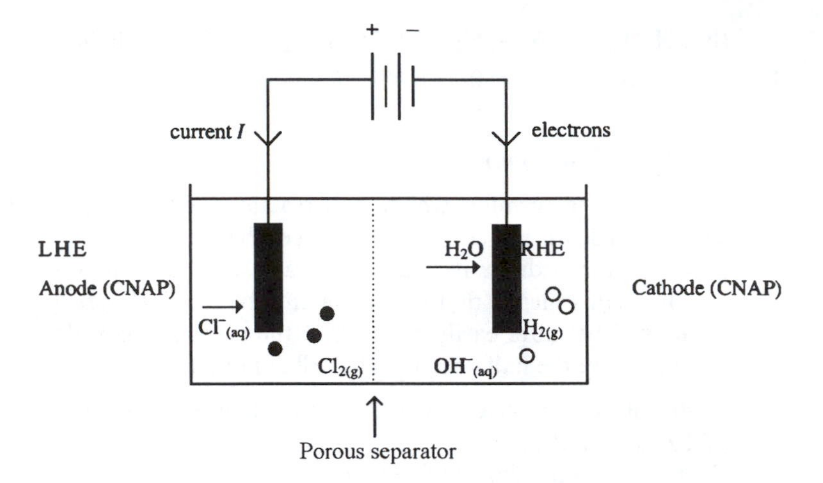

Figure 7.5 *Electrolytic cell for the electrolysis of aqueous sodium chloride*

Although this procedure to determine the respective half-reactions in each case makes logical sense, it is not trivial, and in many problems on electrolytic cells, the standard electrode potentials will not be given, unlike in the analogous galvanic cell problems. However, it should be stated that such half-reactions can quickly be determined, from a knowledge of the Electrochemical Series, and it is not necessary to memorise such half-reactions. Such a criterion leads to a simple rule of thumb:

1. ***Cathode Half-Reactions:*** In the electrolysis of aqueous solutions of salts containing metals above zinc in the electrochemical series, hydrogen gas is generally produced at the cathode, and the half-reaction is: $2H_2O_{(aq)} + 2e \rightarrow H_{2(g)} + 2OH^-_{(aq)}$
2. ***Anode Half-Reactions***: If two or more anions are present in aqueous solution, they are discharged selectively in the following order:

$$I^- > OH^- > Cl^- > NO_3^- > SO_4^{2-}$$

$\longleftarrow$

Ease of Oxidation

This forms the basis of the determination of the half-reactions at both anode and cathode, eliminating the need to examine all possible half-reactions. From this rule of thumb, the case of aqueous sodium chloride is straightforward, *i.e.* although water is present, sodium is above zinc in the electrochemical series, and therefore hydrogen gas is evolved at the cathode. Chloride ions will accumulate at the anode, and chlorine gas will consequently be evolved here.

There is one further factor which must be addressed in these cells, *i.e.* concentration. The electrolysis of hydrochloric acid illustrates this effect, as shown in example (c) below.

Electrolysis of Concentrated Aqueous HCl

1. Determine whether or not the electrolysis involves (a) a molten or (b) an aqueous substance. In this case, the electrolysis involves aqueous hydrochloric acid and therefore additional reactions involving water need to be considered.
2. Identify all species present: $H^+_{(aq)}$, $Cl^-_{(aq)}$ and H_2O.
3. Having identified all the species present, determine which species accumulate at the cathode, and which species accumulate at the anode:

   Cathode $-$ve H^+, H_2O Anode $+$ve Cl^-, H_2O
4. At the cathode, two species appear to be present. However, since hydrochloric acid is a strong acid (*i.e.* a good proton donor, which dissociates readily in solution), it will first ionise in the presence of water to undergo the following reaction: $H_2O + HCl \rightarrow H_3O^+ + Cl^-$ *i.e.* generating H_3O^+ ions (H^+) at the cathode. The cathode reaction then becomes much simpler: $2H^+_{(aq)} + 2e \rightarrow H_{2(g)}$. Hence, hydrogen gas is evolved at the cathode.
   At the anode there are two possible half-reactions:

   (a) $\underset{-I}{Cl^-_{(aq)}} \rightarrow \underset{0}{\tfrac{1}{2}Cl_{2(g)}} + e \qquad E° = -1.36\ V$

   (b) $2\underset{I\ -II}{H_2O_{(aq)}} \rightarrow \underset{0}{O_{2(g)}} + 4\underset{I}{H^+_{(aq)}} + 4e \qquad E° = -1.23\ V$

   -2 to 0 (*oxidation*)

This time, since $E°$ for the chloride is -1.36 V and $E°$ for the water is -1.23 V, the values are so close that it will be concentration dependent as to which half-reaction will actually occur, *i.e.* if the solution is concentrated hydrochloric acid, the concentration of chloride will be large, and chlorine gas will be evolved at the anode, but if the solution is very dilute, then oxygen gas will be evolved at the anode. In this example, let us assume that the acid is very concentrated:

Cathode reaction: $2H^+{}_{(aq)} + 2e^- \rightarrow H_{2(g)}$
Anode reaction: $2Cl^-{}_{(aq)} \rightarrow Cl_{2(g)} + 2e$

Cell reaction: $2H^+{}_{(aq)} + 2Cl^-{}_{(aq)} \rightarrow Cl_{2(g)} + H_{2(g)}$

The electrolysis of hydrochloric acid is very important in stressing the role of the initial acid ionisation reaction and the effect of concentration in these types of electrolytic cells.

5. Draw the cell (Figure 7.6).

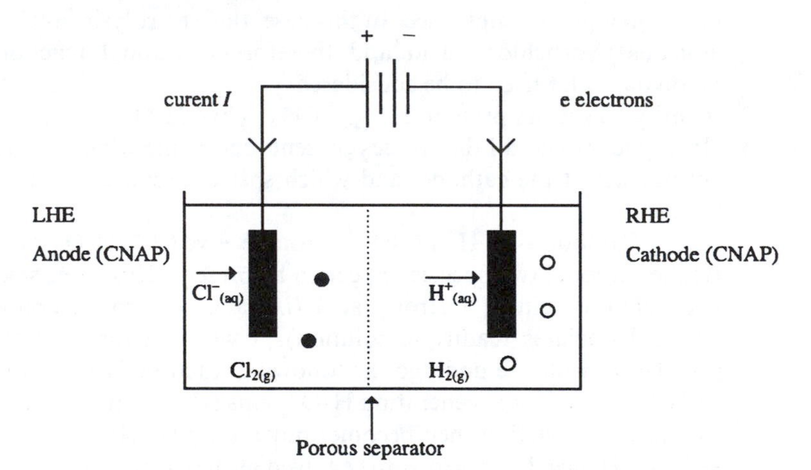

Figure 7.6 *Electrolytic cell for the electrolysis of concentrated aqueous hydrochloric acid*

Electrolysis of Aqueous H_2SO_4

1. Aqueous solution, so the additional factor of H_2O has to be considered.
2. Identify all species present in solution: $H^+{}_{(aq)}$, $SO_4{}^{2-}{}_{(aq)}$ and H_2O.
3. Cathode −ve H^+, H_2O Anode +ve $SO_4{}^{2-}$, H_2O
4. Sulfuric acid is a strong diprotic acid (*i.e.* contains two replaceable hydrogens) which will first ionise in the presence of water to undergo the following reactions:

$$H_2O + H_2SO_4 \rightarrow HSO_4{}^- + H_3O^+$$
$$H_2O + HSO_4{}^- \rightarrow SO_4{}^{2-} + H_3O^+$$
$$i.e.\quad H_2SO_{4(aq)} \rightarrow 2H^+{}_{(aq)} + SO_4{}^{2-}{}_{(aq)}$$

i.e. generating H^+ ions at the cathode. The leads to only one posssible reaction at the cathode: $2H^+_{(aq)} + 2e \rightarrow H_{2(g)}$.
At the anode, there are two possible half-reactions:

(a) $2SO_4{}^{2-}{}_{(aq)} \rightarrow S_2O_8{}^{2-}{}_{(aq)} + 2e \qquad E^\circ = -2.05$ V.
S^{VI} Peroxydisulfate anion
This is a very unfavourable half-reaction (as shown by a large −ve E° value, *i.e.* ΔG° very +ve, implying a non-spontaneous reaction). Sulfur is not oxidised, as in the VI oxidation state, it has the stable inert gas core configuration of [Ne], *i.e.* oxidation of the sulfate will not occur, and water will instead be oxidised:

(b) $2H_2O_{(aq)} \rightarrow O_{2(g)} + 4H^+_{(aq)} + 4e \qquad E^\circ = -1.23$ V.
I −II 0 I

−2 to 0 (*oxidation*)

Cathode reaction: $2H^+_{(aq)} + 2e \rightarrow H_{2(g)}$
Anode reaction: $2H_2O_{(aq)} \rightarrow O_{2(g)} + 4H^+_{(aq)} + 4e$

Multiply by two: Cathode reaction: $4H^+_{(aq)} + 4e \rightarrow 2H_{2(g)}$
Anode reaction: $2H_2O_{(aq)} \rightarrow O_{2(g)} + 4H^+_{(aq)} + 4e$

Cell reaction: $2H_2O_{(aq)} \rightarrow 2H_{2(g)} + O_{2(g)}$

5. Draw the cell (Figure 7.7).

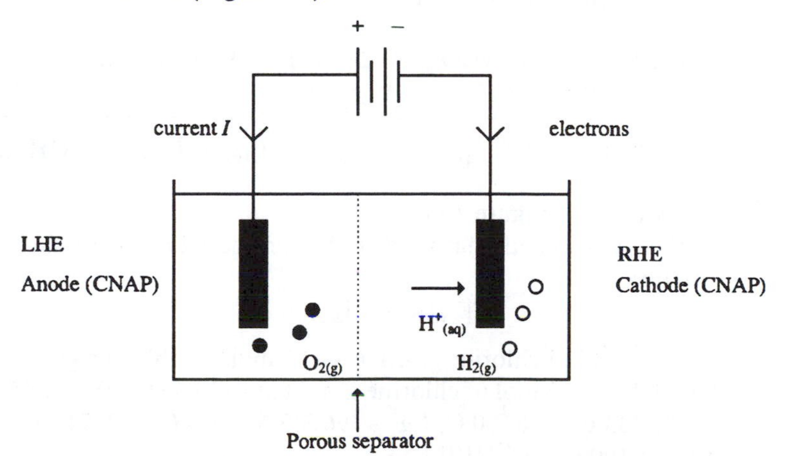

Figure 7.7 *Electrolytic cell for the electrolysis of aqueous sulfuric acid*

Therefore, hydrogen gas is evolved at the cathode, and oxygen gas is evolved at the anode in the electrolysis of aqueous sulfuric acid.

EXAMPLES OF ELECTROLYTIC CELL TYPE PROBLEMS

Example No. 1: Chlorine is produced by the electrolysis of aqueous sodium chloride. Draw the electrochemical cell, indicating clearly the anode, the cathode, the direction of electron flow and the direction of the current. Assuming that chlorine is the only species produced at the anode, determine how long it will take to produce 1 kg of chlorine gas, in a cell operating at 950 A. (F = 96 500 C mol^{-1}; molar mass (M) of Cl = 35.453 g mol^{-1}).

Solution:

1. Type of cell: electrolytic.
2. Type of electrolysis: electrolysis of aqueous sodium chloride. Hence, water must be considered.
3. Identify all species present: $Na^{+}_{(aq)}$, $Cl^{-}_{(aq)}$, H_2O.
4. Cathode: −ve electrode ('CNAP'): $Na^{+}_{(aq)}$, H_2O.
   Na is above Zn in the electrochemical series; therefore $H_{2(g)}$ is discharged at the cathode, according to the half-reaction: $2H_2O_{(aq)} + 2e \rightarrow H_{2(g)} + 2OH^{-}_{(aq)}$.
5. Anode: +ve electrode ('CNAP'): $Cl^{-}_{(aq)}$, H_2O.
   In this question, it is stated that chlorine gas is evolved at the anode. Hence, $2Cl^{-}_{(aq)} \rightarrow Cl_{2(g)} + 2e$.
6. Write down the two half-reactions:

   Cathode reaction: $2H_2O_{(aq)} + 2e \rightarrow H_{2(g)} + 2OH^{-}_{(aq)}$
   Anode reaction: $2Cl^{-}_{(aq)} \rightarrow Cl_{2(g)} + 2e$

   Cell reaction: $2H_2O_{(aq)} + 2Cl^{-}_{(aq)} \rightarrow H_{2(g)} + Cl_{2(g)} + 2OH^{-}_{(aq)}$

7. Draw the cell (Figure 7.8).
8. In this question, all that needs to be considered is the anode half-reaction:

   $$2Cl^{-}_{(aq)} \rightarrow Cl_{2(g)} + 2e$$

   $2\,F \rightarrow$ 1 mol of chlorine gas; $1\,F \rightarrow$ 0.5 mole of chlorine gas
   96 500 C $\rightarrow$ 0.5 mol of chlorine gas = 0.5 × (35.453 × 2) = 35.453 g
   i.e. 35.453 g $\rightarrow$ 96 500 C; 1 g $\rightarrow$ (96 500/35.453) C = 2721.9135 C
   1 kg = 1000 g $\rightarrow$ 2721913.5 C
9. $Q = It \Rightarrow t = Q/I$ = 2721913.5/950 = 2865.172105 s
   = (2865.172105/60) min = 47.75 min.

Answer: 47.75 min

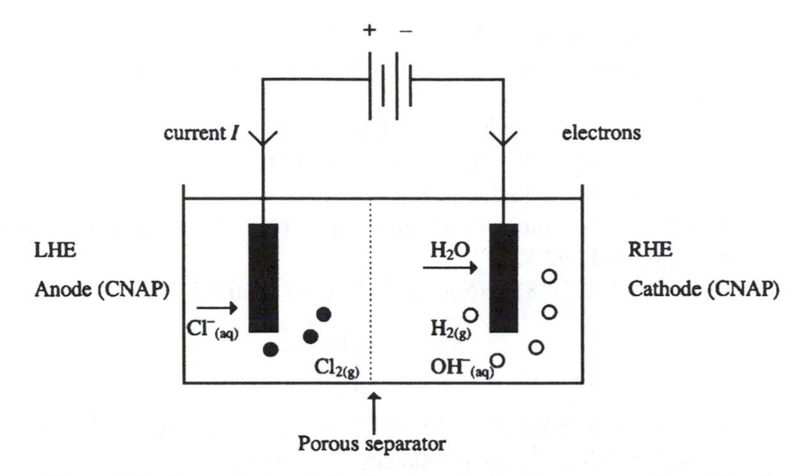

Figure 7.8 *Electrolytic cell for the electrolysis of aqueous sodium chloride*

> ***Example No. 2:*** Estimate the rate of chlorine gas evolution (in $cm^3\ min^{-1}$) at the anode of an aqueous sodium chloride electrolysis cell, operating at a current of 650 mA, T = 300 K and p = 1.2 bar. Draw the electrochemical cell, indicating clearly the anode, the cathode, the direction of electron flow and the direction of current. (F = 96 500 C mol^{-1}; 1 mol of an ideal gas at 25 °C and 1 bar pressure occupies 24.8 dm^3).

Solution:

1. Type of cell: electrolytic.
2. Type of electrolysis: electrolysis of aqueous sodium chloride. Hence, water must also be considered.
3. Identify all species present: $Na^+{}_{(aq)}$, $Cl^-{}_{(aq)}$, H_2O.
4. Cathode: −ve electrode ('CNAP'): $Na^+{}_{(aq)}$, H_2O.
   Na is above Zn in the electrochemical series; therefore $H_{2(g)}$ is discharged at the cathode according to the half-reaction: $2H_2O_{(aq)} + 2e \rightarrow H_{2(g)} + 2OH^-{}_{(aq)}$.
5. Anode: +ve electrode ('CNAP'): $Cl^-{}_{(aq)}$, H_2O.
   Chlorine gas is evolved at the anode (stated in question). Hence $2Cl^-{}_{(aq)} \rightarrow Cl_{2(g)} + 2e$.
6. Write down the two half-reactions:

   Cathode reaction: $2H_2O_{(aq)} + 2e \rightarrow H_{2(g)} + 2OH^-{}_{(aq)}$
   Anode reaction: $2Cl^-{}_{(aq)} \rightarrow Cl_{2(g)} + 2e$

   Cell reaction: $2H_2O_{(aq)} + 2Cl^-{}_{(aq)} \rightarrow H_{2(g)} + Cl_{2(g)} + 2OH^-{}_{(aq)}$

7. Draw the cell (Figure 7.8).
8. In this question, all that has to be considered is the anode half-reaction:

$$2Cl^-_{(aq)} \rightarrow Cl_{2(g)} + 2e$$

Per minute, $Q = It = 0.650 \times 60 = 39$ C;
2 $F \rightarrow$ 1 mol of chlorine gas
96 500 C $\rightarrow$ 0.5 mol of chlorine gas $= (0.5 \times 24.8\ dm^3)$ at 1 bar pressure and 298 K
1 C $\rightarrow (0.5 \times 24.8)/96\,500\ dm^3$; 39 C $\rightarrow 0.0050114\ dm^3$

$$\frac{p_1 V_1}{T_1} = \frac{p_2 V_2}{T_2}$$

$V_2 = (1 \times 0.0050114 \times 300)/(1.2 \times 298) = 0.0042\ dm^3\ min^{-1}$
i.e. $4.2\ cm^3\ min^{-1}$, since $1000\ cm^3 = 1\ dm^3$.

Answer = 4.2 cm³ min⁻¹

> ***Example No. 3:*** A slightly acidified solution of copper(II) sulfate is electrolysed using inert platinum electrodes. Oxygen gas is evolved at one electrode, and copper metal is deposited at the other. Draw a fully labelled diagram of the cell, indicating clearly the cathode, the anode, the direction of current, the direction of electrons and ion flow. Write down the cathode, anode and cell reactions. How long would it take this cell to deliver 2 dm^3 of $O_{2(g)}$ at 1.5 bar pressure and 65 °C, using an electrolysis current of 475 mA? ($F = 96\,500\ C\ mol^{-1}$; 1 mol of an ideal gas at 25 °C and 1 bar pressure occupies 24.8 dm^3).

Solution:

1. Type of cell: electrolytic.
2. Type of electrolysis: electrolysis of aqueous copper(II) sulfate.
3. Identify all species present: $Cu^{2+}_{(aq)}$, $SO_4^{2-}{}_{(aq)}$, H_2O.
4. Cathode: −ve electrode ('CNAP'): $Cu^{2+}_{(aq)}$, H_2O.
   Cu is below Zn in the electrochemical series; therefore copper metal is deposited at the cathode:

$$Cu^{2+}_{(aq)} + 2e \rightarrow Cu^0_{(s)}$$

5. Anode: +ve electrode ('CNAP'): $SO_4^{2-}{}_{(aq)}$, H_2O. But $SO_4^{2-}{}_{(aq)}$ is not easily oxidised:

$$I^- > OH^- > Cl^- > NO_3^- > SO_4^{2-}$$

$$\overleftarrow{\hspace{6cm}}$$

Ease of Oxidation

Therefore water is oxidised at the anode, evolving oxygen gas, according to the half-reaction: $2H_2O_{(aq)} \rightarrow O_{2(g)} + 4H^+_{(aq)} + 4e$.

6. Write down the two half-reactions:

$$\text{Cathode reaction: } Cu^{2+}_{(aq)} + 2e \rightarrow Cu^0_{(s)}$$
$$\text{Anode reaction: } 2H_2O_{(aq)} \rightarrow O_{2(g)} + 4H^+_{(aq)} + 4e$$

Multiply by two:
$$\text{Cathode reaction: } 2Cu^{2+}_{(aq)} + 4e \rightarrow 2Cu^0_{(s)}$$
$$\text{Anode reaction: } 2H_2O_{(aq)} \rightarrow O_{2(g)} + 4H^+_{(aq)} + 4e$$

$$\text{Cell reaction: } 2H_2O_{(aq)} + 2Cu^{2+}_{(aq)} \rightarrow 2Cu_{(s)} + O_{2(g)} + 4H^+_{(aq)}$$

7. Draw the cell (Figure 7.9).

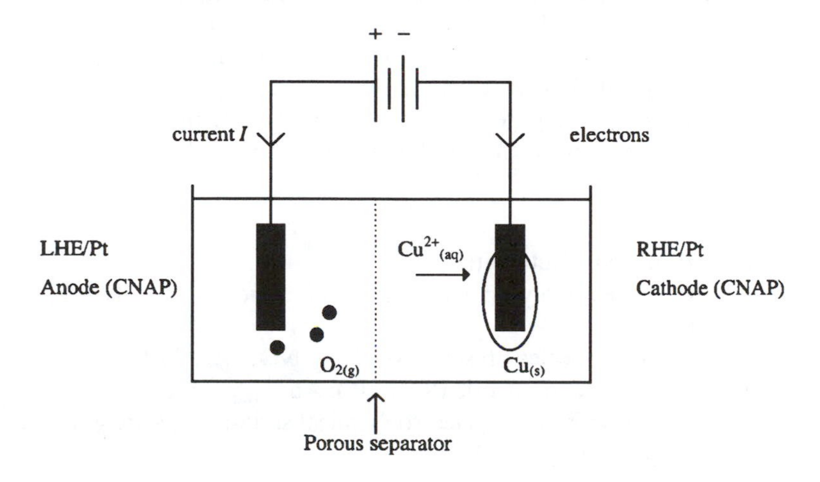

Figure 7.9 *Electrolytic cell for the electrolysis of aqueous copper(II) sulfate*

8. Consider the anode half-reaction: $2H_2O_{(aq)} \rightarrow O_{2(g)} + 4H^+_{(aq)} + 4e$

$$Q = It; \ t = Q/I$$

Therefore, since $I = 475$ mA $= 0.475$ A, Q must be determined at 65 °C and 1.5 bar pressure.

$$\frac{p_1 V_1}{T_1} = \frac{p_2 V_2}{T_2} \quad \text{must be applied.}$$

First, convert T_2 to K: $T_2 = (273 + 65) = 338$ K.
$V_2 = (1.5 \times 2 \times 298)/(1 \times 338) = 2.645$ dm^3.

9. Apply Faraday's Second Law of Electrolysis:

$$2H_2O_{(aq)} \rightarrow O_{2(g)} + 4H^+_{(aq)} + 4e$$

$4\ F \rightarrow 1$ mol of oxygen; 1 mol of oxygen $\rightarrow 4 \times 96\,500$ C; $24.8\ dm^3 \rightarrow 4 \times 96\,500$ C; $1\ dm^3 \rightarrow (4 \times 96\,500)/24.8 = 15564.52$ C; $2.645\ dm^3 \rightarrow 41168.16$ C.
Now $t = Q/I = 41168.16/0.475 = 86669.81$ s $= 24.07$ h.

Answer: 24.07 h

Example No. 4: An aqueous solution of gold(III) nitrate, $Au(NO_3)_3$, was electrolysed with a current of 219 mA, until 150 cm^3 of oxygen gas was liberated at the anode, at 1 bar pressure and 298 K. Draw a fully labelled diagram of the cell, indicating clearly the cathode, the anode, the direction of current, the direction of electrons and ion flow. Write down the cathode, anode and cell reactions. Find (a) the duration of the experiment and (b) the mass of gold deposited at the cathode during electrolysis. ($F = 96\,500$ C mol^{-1}; 1 mole of an ideal gas at 25 °C and 1 bar pressure occupies 24.8 dm^3; molar mass (M) of Au = 196.97 g mol^{-1}).

Solution:

1. Type of cell: electrolytic.
2. Type of electrolysis: electrolysis of aqueous gold(III) nitrate, $Au(NO_3)_3$.
3. Identify all species present: $Au^{3+}_{(aq)}$, $NO_3^-{}_{(aq)}$, H_2O.
4. Cathode: −ve electrode ('CNAP'): $Au^{3+}_{(aq)}$, H_2O.
   Au is below Zn in the electrochemical series; therefore gold metal is deposited at the cathode:

$$Au^{3+}_{(aq)} + 3e \rightarrow Au_{(s)}$$

5. Anode: +ve electrode ('CNAP'): $NO_3^-{}_{(aq)}$, H_2O. But $NO_3^-{}_{(aq)}$ is not easily oxidised according to its position in the electrochemical series:

$$I^- > OH^- > Cl^- > NO_3^- > SO_4^{2-}$$

$$\parallel$$

$$\longleftarrow$$

Ease of Oxidation

   Therefore water is oxidised at the anode, evolving oxygen gas, according to the half-reaction: $2H_2O_{(aq)} \rightarrow O_{2(g)} + 4H^+_{(aq)} + 4e$
6. Write down the two half-reactions:

Cathode reaction: $Au^{3+}_{(aq)} + 3e \rightarrow Au^{0}_{(s)}$
Anode reaction: $2H_2O_{(aq)} \rightarrow O_{2(g)} + 4H^{+}_{(aq)} + 4e$

Multiply by four: Cathode reaction: $4Au^{3+}_{(aq)} + 12e \rightarrow 4Au^{0}_{(s)}$
Multiply by three: Anode reaction: $6H_2O_{(aq)} \rightarrow 3O_{2(g)} + 12H^{+}_{(aq)} + 12e$

Cell reaction: $6H_2O_{(aq)} + 4Au^{3+}_{(aq)} \rightarrow 4Au^{0}_{(s)} + 3O_{2(g)} + 12H^{+}_{(aq)}$

7. Draw the cell (Figure 7.10).

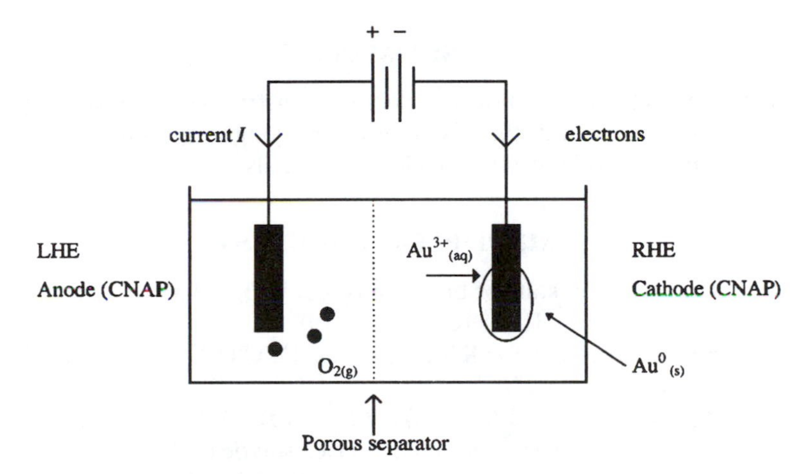

Figure 7.10 *Electrolytic cell for the electrolysis of aqueous gold(III) nitrate*

8. $Q = It$; $t = Q/I$. Therefore, since $I = 219$ mA $= 0.219$ A, Q needs to be determined.
9. Apply Faraday's Second Law of Electrolysis:

$$6H_2O_{(aq)} \rightarrow 3O_{2(g)} + 12H^{+}_{(aq)} + 12e$$

12 $F \rightarrow$ 3 mol of oxygen; 3 mol of oxygen $\rightarrow 12 \times 96\,500$ C; 1 mol of oxygen $\rightarrow 4 \times 96\,500$ C (†); 24.8 dm$^3 \rightarrow 4 \times 96\,500$ C; 1 dm$^3 \rightarrow (4 \times 96\,500/24.8) = 15564.52$ C. But 1 dm^3 = 1 000 cm^3 $\Rightarrow$ 150 cm^3 = 0.150 dm^3; 0.150 dm$^3 \rightarrow 15564.52 \times 0.150 =$ 2334.678 C.
Now $t = Q/I = 2334.678/0.219 = 10660.63014$ s = 2.96 h.

Answer: 2.96 h

10. Apply Faraday's Second Law again:

$$6H_2O_{(aq)} + 4Au^{3+}{}_{(aq)} \rightarrow 4Au^0{}_{(s)} + 3O_{2(g)} + 12H^+{}_{(aq)}$$

3 mol of $O_{2(g)}$ → 4 mol of Au; 1 mol of $O_{2(g)}$ → (4/3) mol of Au; 4 × 96 500 C → (4/3) mol of Au (†); 1 C → (4/3)/(4 × 96 500) mol of Au = 3.45×10^{-6} mol of Au; 2334.678 C → 0.00805464 mol = 1.59 g of Au.

Answer: 1.59 g

SUMMARY

This completes the section on electrochemistry. The following multiple-choice test and the six subsequent longer questions contain questions on both galvanic and electrolytic cells.

MULTIPLE-CHOICE TEST

1. The oxidation state of boron in $Na_2BeB_4O_8 \cdot 7H_2O$ is:
   (a) I (b) II (c) III (d) IV
2. For the reaction ν_A $KMnO_4$ + ν_B $Fe(ClO_3)_3$ →, the stoichiometry ratio $\nu_A:\nu_B$ is:
   (a) 6:5 (b) 5:6 (c) 7:6 (d) 6:7
3. What energy change occurs in an electrolytic cell?
   (a) chemical to electrical; (b) electrical to chemical; (c) both electrical to chemical and chemical to electrical; (d) you cannot say for sure—such changes are concentration dependent.
4. For the electrochemical cell $Zn|Zn^{2+}||Fe^{3+}, Fe^{2+}|Pt$, which of the following statements is correct, given that $E°(Zn^{2+},Zn) = -1.18$ V and $E°(Fe^{3+}, Fe^{2+}) = +0.44$ V?
   (a) Both electrodes are metal–metal-ion electrodes; (b) The $Zn|Zn^{2+}$ electrode is the cathode; (c) $E°_{cell} = -0.74$ V; (d) $E°_{cell} = +1.62$ V.
5. When 1.48 A of current is passed through a solution of $ZnSO_4$ for 1.5 hours, the mass in g, of Zn deposited at the cathode is:
   (a) 0.045 (b) 5.42 (c) 0.00075 (d) 2.71
   Molar mass (M) of Zn = 65.39 g mol^{-1} and F = 96 500 C mol^{-1}.
6. Determine $\Delta G°$ in kJ mol^{-1} at 25 °C for the following reaction:

$$Sn^{2+}{}_{(aq)} + 2Fe^{3+}{}_{(aq)} \rightarrow Sn^{4+}{}_{(aq)} + 2Fe^{2+}{}_{(aq)}$$

$E°(Sn^{4+}, Sn^{2+}) = 0.15$ V and $E°(Fe^{3+}, Fe^{2+}) = 0.77$ V and $F = 96\,500$ C mol^{-1}.

(a) +60 (b) 120 (c) −60 (d) −120

7. In the electrolysis of molten sodium chloride, how many of the following statements are incorrect?
   * Only sodium cations are reduced;
   * Chlorine gas is evolved at the cathode;
   * Hydrogen gas is generated at the anode;
   * Chlorine gas is evolved at the anode.

   (a) All 4 (b) 3 (c) 2 (d) 1
8. The volume of hydrogen gas in dm^3 measured at 1 bar and 298 K, evolved by a current of 8 mA flowing for a period of 25 days, through a concentrated aqueous hydrochloric acid electrolysis cell is:

   (a) 2.22 (b) 4.44 (c) 214272 (d) 2220

   ($F = 96\,500$ C mol^{-1} and 1 mol of an ideal gas at 1 bar pressure and 298 K occupies 24.8 dm^3).
9. How many of the following equations are incorrect?

   $Q = ItE°$; $E°_{\text{cell}} = E°_{\text{LHE}} - E°_{\text{RHE}}$; $m = zIT$; $\Delta G = -nFE$;

   (a) All 4 (b) 3 (c) 2 (d) 1
10. How many of the following electrodes will cause hydrogen gas to be generated when they are coupled with a hydrogen electrode to form an electrochemical cell?

    $E°(Fe^{2+}, Fe) = -0.44$ V; $E°(Ag^{+}, Ag) = +0.80$ V;
    $E°(Ca^{2+}, Ca) = -2.87$ V; $E°(Sn^{4+}, Sn^{2+}) = 0.15$ V.

    (a) 4 (b) 3 (c) 2 (d) 1

LONG QUESTIONS ON CHAPTERS 6 AND 7

1. Give a fully labelled diagram of a galvanic cell based on the following reaction:

   $$PbSO_{4(s)} + 2FeSO_{4(aq)} \rightarrow Fe_2(SO_4)_{3(aq)} + Pb_{(s)}$$

   If the cell is short-circuited, indicate the direction of current, electron flow and ion flow. Label clearly the anode, the cathode, and give the electrode and cell reactions. Given that $E°(Pb|PbSO_4|SO_4{}^{2-}) = -0.36$ V, $E°(Fe^{3+}, Fe^{2+}) = +0.77$ V and $F = 96\,500$ C mol^{-1}, comment on the spontaneity of the cell. Explain your answer, by calculating the value of $\Delta G°$.
2. Draw the Daniel cell. Determine the potential of the cell at 25 °C, in which the molar concentration of the Zn^{2+} ions is 0.22 M and that of the Cu^{2+} ions is 0.003 M, given that

$E°(Zn^{2+} \mid Zn) = -0.76$ V and $E°(Cu^{2+} \mid Cu) = +0.34$ V. (R = 8.314 J K^{-1} mol^{-1}; F = 96 500 C mol^{-1}).

3. Calculate the equilibrium constant of the following reaction at 300 K:

$$2Fe^{2+}{}_{(aq)} + Cl_{2(g)} \rightleftharpoons 2Fe^{3+}{}_{(aq)} + 2Cl^{-}{}_{(aq)}$$

Sketch the cell that would have this as its cell reaction. If the cell is short-circuited, indicate clearly the cathode, the anode, the direction of current, the direction of electrons and the direction of ion flow, given that $E°(Fe^{3+}, Fe^{2+}) = +0.77$ V, $E°(Cl_2, Cl^-)$ = 1.36 V, R = 8.314 J K^{-1} mol^{-1} and F = 96 500 C mol^{-1}). Comment on the value of the equilibrium constant.

4. Calculate the volume (in cm^3) of hydrogen gas produced at 25 °C and 0.65 bar by the electrolysis of water when 0.045 mol of electrons is supplied. (1 mole of an ideal gas at 1 bar pressure and 298 K occupies 24.8 dm^3.)

5. Chlorine gas is produced by the electrolysis of concentrated aqueous sodium chloride. Give a diagram of this process, in particular, illustrating the electron flow and the conventional direction of current. Assuming an anode efficiency of 100%, how long will it take to produce 0.0025 g of chlorine, in a cell operating at 760 μA? (F=96 500 C mol^{-1} and molar mass (M) of Cl=35.453 g mol^{-1}.)

6. The same quantity of electricity that caused the deposition of 10.2 g of Ag from an $AgNO_3$ solution caused the discharge of 6.11 g of Au when passed through a solution containing gold cations of unknown charge, Au^{m+}. Determine m, given that the molar masses (M) of Ag and Au are 107.87 and 196.97 g mol^{-1} respectively, and F = 96 500 C mol^{-1}.

Chapter 8

Chemical Kinetics I: Basic Kinetic Laws

RATE OF A REACTION

Any chemical reaction may be represented by the stoichiometric equation

$$\nu_A A + \nu_B B + \ldots \rightarrow \nu_C C + \nu_D D + \ldots$$

where ν_A, ν_B, ν_C and ν_D are termed the ***stoichiometry factors***. Chemical kinetics is concerned with the rate at which reactions proceed. The rate of a chemical reaction is a measure of how fast the products (C, D, . . .) are formed and how fast the reactants (A, B, . . .) are consumed. The rate of a chemical reaction is related to the stoichiometry. Consider the reaction:

$$A + 4B \rightarrow 3C + 2D$$

where $\nu_A = 1$, $\nu_B = 4$, $\nu_C = 3$ and $\nu_D = 2$.

$$-\frac{d[A]}{dt} = \text{rate of decrease of } [A]$$

$$-\frac{d[B]}{dt} = \text{rate of decrease of } [B]$$

$$+\frac{d[C]}{dt} = \text{rate of increase of } [C]$$

$$+\frac{d[D]}{dt} = \text{rate of increase of } [D]$$

where the rate is proportional to the reciprocal of time, *i.e.* 1/time or d/dt expressed as a derivative.

However, an examination of the equation for the reaction shows that the rates of these four changes are not equivalent, but are related to ν_A, ν_B, ν_C and ν_D respectively, *i.e.* the stoichiometry factors of the reaction. It can be seen in the above example, that B is consumed four times as fast as A, *i.e.* 1 mole of A consumes 4 moles of B, and this in turn is related to the rate of formation of C and D respectively, *i.e.*:

$$-\frac{1}{1}\frac{d[A]}{dt} = -\frac{1}{4}\frac{d[B]}{dt} = +\frac{1}{3}\frac{d[C]}{dt} = +\frac{1}{2}\frac{d[D]}{dt}$$

In general, the rate of a chemical reaction:

$$\nu_A A + \nu_B B + \ldots \rightarrow \nu_C C + \nu_D D + \ldots$$

can be expressed as:

$$-\frac{1}{\nu_A}\frac{d[A]}{dt} = -\frac{1}{\nu_B}\frac{d[B]}{dt} = +\frac{1}{\nu_C}\frac{d[C]}{dt} = +\frac{1}{\nu_D}\frac{d[D]}{dt}$$

or $\pm\frac{1}{\nu_J}\frac{d[J]}{dt}$ as a general formula.

RATE LAW

This is the algebraic statement of the dependence of a chemical reaction on the concentrations of a number of species, normally the reactants. Consider the reaction:

$$A + B \rightarrow C$$

where $\nu_A = 1$, $\nu_B = 1$ and $\nu_C = 1$.

$$\text{Rate} = -\frac{1}{1}\frac{d[A]}{dt} = -\frac{1}{1}\frac{d[B]}{dt} = +\frac{1}{1}\frac{d[C]}{dt}$$

where $-d[A]/dt \propto [A]$ and $-d[A]/dt \propto [B]$

$$\text{Rate} = -\frac{1}{1}\frac{d[A]}{dt} \propto [A][B]$$

$$\text{Rate} = -\frac{1}{1}\frac{d[A]}{dt} = k[A][B]$$

i.e. Rate $= k[A][B]$, where k, the constant of proportionality, is known as the ***specific rate constant***. Such an equation is an example of the rate law of a reaction, or more specifically, the experimental rate equation.

ORDER OF A REACTION

The ***order*** of a reaction is the sum of the exponents on the concentration terms in the rate equation, *i.e.* $[\ ]^n$, where n = the order of the reactant in question. In this text the abbreviations 1°, 2°, 3° *etc.* will occasionally be used for simplicity.

Consider a reaction, such that: Rate = $k[A]^3[B]^1$

This reaction is: (a) third-order in A; (b) first-order in [B] and (c) (3 + 1) = fourth-order overall.

Orders do not necessarily have to be whole numbers (*e.g.* −1, −2, +3, +1 *etc.*), but can also take fractional values *e.g.* $\frac{1}{2}$, $\frac{1}{3}$ *etc.*), *i.e.* orders can be intermediate, or even zero.

Zero-Order Reactions

Consider a reaction such that A → Products

$$\text{Rate} = -\frac{1}{1}\frac{d[A]}{dt} = k[A]^0 = k$$

since $x^0 = 1$, from the rules of indices

$$\Rightarrow \quad -\frac{d[A]}{dt} = k$$

$$\Rightarrow \quad -d[A] = kdt$$

$$\Rightarrow \quad -\int d[A] = k\int dt$$

$$\Rightarrow \quad -[A] = kt + c$$

At $t = 0$, $[A] = [A_0]$ = the initial concentration $\Rightarrow c = -[A_0]$

$$\Rightarrow \quad -[A] = kt + (-[A_0])$$

$$\boxed{\begin{array}{lllll} [A] & = & -kt & + & [A_0] \\ y & = & mx & + & c \text{ (zero-order reaction)} \end{array}}$$

This is a linear equation, which is the equation of a line, where m = the slope or gradient = $(y_2 - y_1)/(x_2 - x_1)$ = $\Delta y/\Delta x$, *i.e.* (*difference of the y's*)/(*difference of the x's*), and c = the intercept, *i.e.* the point where the graph cuts the y-axis, at $x = 0$ (Figure 8.1).

Units of k for a zero-order reaction:

$$-kt = [A] - [A_0] \quad \Rightarrow kt = [A_0] - [A]$$

$$\Rightarrow k = [A_0] - [A]/t \quad \Rightarrow \text{ units of } k \text{ are M s}^{-1}.$$

It should be noted that zero-order reactions are rare, usually found in the case of surface reactions. First-order reactions are much more common, found for species in the gaseous or aqueous phase.

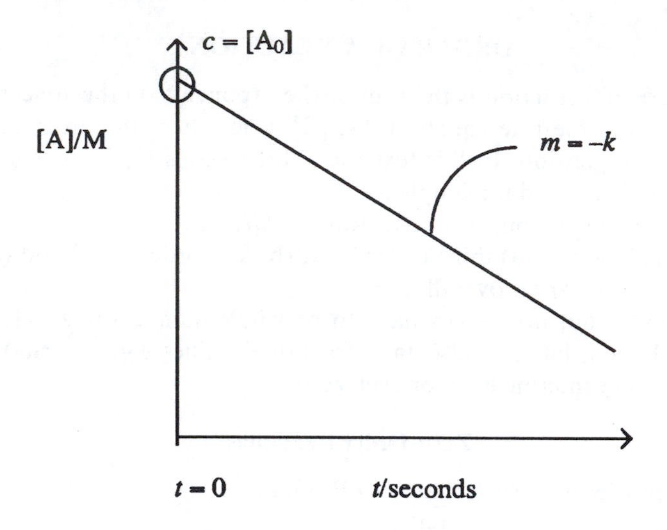

Figure 8.1 *Straight-line graph of a zero-order reaction*

First-Order Reactions

A → Products

$$
\begin{aligned}
\text{Rate} = -\frac{1}{1}\frac{\mathrm{d}[\mathrm{A}]}{\mathrm{d}t} &= k[\mathrm{A}]^1 \\
\Rightarrow \quad -\frac{\mathrm{d}[\mathrm{A}]}{[\mathrm{A}]} &= k\mathrm{d}t \\
\Rightarrow \quad -\int \frac{\mathrm{d}[\mathrm{A}]}{[\mathrm{A}]} &= k\int \mathrm{d}t \\
\Rightarrow \quad -\ln\,[\mathrm{A}] &= kt + c
\end{aligned}
$$

since $\int \frac{1}{x}\mathrm{d}x = \ln|x|$

At $t = 0$, $[\mathrm{A}] = [\mathrm{A}_0]$ = the initial concentration $\Rightarrow c = -\ln[\mathrm{A}_0]$

$$
\begin{aligned}
\Rightarrow \quad -\ln[\mathrm{A}] &= kt - \ln[\mathrm{A}_0] \\
\Rightarrow \quad \ln[\mathrm{A}] &= -kt + \ln[\mathrm{A}_0]
\end{aligned}
$$

$$
\boxed{\begin{aligned}
\ln[\mathrm{A}] &= -kt + \ln[\mathrm{A}_0] \\
y &= mx + c \text{ (first-order reaction)}
\end{aligned}}
$$

This is also the equation of a line, *i.e.* if the reaction obeys first-order kinetics (Figure 8.2), a plot of ln[A] *versus* t should yield a straight line graph of slope $m = -k$ and intercept $c = \ln[A_0]$. *Remember, logs have no units!*

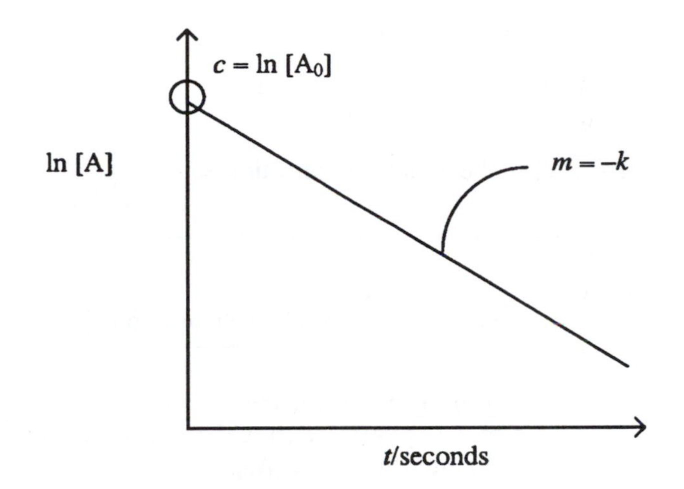

Figure 8.2 *Straight-line graph of a first-order reaction*

Units of k for a first-order reaction:

$$-kt = \ln[A] - \ln[A_0] \Rightarrow kt = \ln[A_0] - \ln[A]$$
$$\Rightarrow k = \{\ln[A_0] - \ln[A]\}/t; \Rightarrow \text{units of } k \text{ are s}^{-1}.$$

Although the units of concentration are mol dm^{-3} (M), logarithmic values are always dimensionless.

Second-Order Reactions

The equations that relate the concentrations of reactants and the rate constant, k, for second-order reactions are much more complicated, and will not be presented in detail in this book, as this is beyond the scope of this introductory course in physical chemistry. However, the equation involving the second-order reaction with respect to one reactant, *i.e.* A $\rightarrow$ Products, will be given:

$$\text{Rate} = -\frac{1}{1}\frac{d[A]}{dt} = k[A]^2$$

$$\Rightarrow -\frac{d[A]}{[A]^2} = k dt$$

$$\Rightarrow -\int\frac{d[A]}{[A]^2} = k\int dt$$

$$\Rightarrow \frac{1}{[A]} = kt + c$$

since $\int x^n dx = \frac{x^{n+1}}{n+1}$

At $t = 0$, $[A] = [A_0]$ = the initial concentration $\Rightarrow c = \frac{1}{[A_0]}$

$$\frac{1}{[A]} = kt + \frac{1}{[A_0]}$$
$$y = mx + c \text{ (second-order reaction)}$$

This is also the equation of a line, *i.e.* if the reaction obeys second-order kinetics, a plot of $1/[A]$ *versus* t should yield a straight line graph, of slope $m = k$ and intercept $c = 1/[A_0]$.

Units of k for a second-order reaction of the form A → Products:

$$\frac{1}{[A]} = kt + \frac{1}{[A_0]}$$

$$\Rightarrow kt = \frac{1}{[A]} - \frac{1}{[A_0]}$$

$$\Rightarrow k = \left\{\frac{1}{[A]} - \frac{1}{[A]}\right\}/t$$

$\Rightarrow$ units of k are $M^{-1}s^{-1}$.

HALF-LIFE

The ***half-life*** is the time it takes one-half of a reactant to be consumed in a reaction (Figure 8.3).

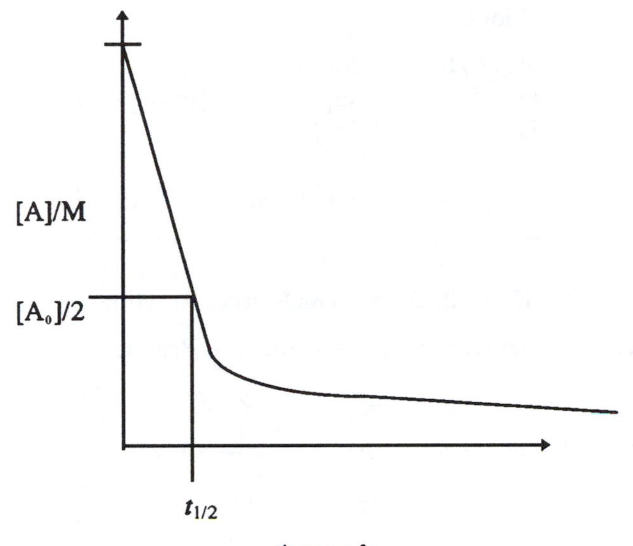

Figure 8.3 *Half-life of a reaction*

Half-Life for a Zero-Order Reaction

For a zero-order reaction:

$$
\begin{aligned}
& [A] &= -kt + [A_0] \\
\text{When } t = t_{1/2}, \quad & [A] &= [A_0]/2 \\
\Rightarrow \quad & [A_0]/2 &= -kt_{1/2} + [A_0] \\
\Rightarrow \quad & kt_{1/2} &= [A_0] - [A_0]/2 \\
\Rightarrow \quad & kt_{1/2} &= [A_0]/2
\end{aligned}
$$

$$t_{1/2} = \frac{[A_0]}{2k} \qquad \text{(zero-order reaction)}$$

Half-Life for a First-Order Reaction

For a first-order reaction:

$$
\begin{aligned}
& \ln[A] &= -kt + \ln[A_0] \\
\text{When } t = t_{1/2}, \quad & [A] &= [A_0]/2 \\
\Rightarrow \quad & \ln\{[A_0]/2\} &= -kt_{1/2} + \ln[A_0] \\
\Rightarrow \quad & kt_{1/2} &= \ln[A_0] - \ln\{[A_0]/2\}
\end{aligned}
$$

From the rules of logs

$$\begin{aligned} \log(A/B) &= \log A - \log B \\ \Rightarrow kt_{1/2} &= \ln[A_0] - \{\ln[A_0] - \ln 2\} \\ \Rightarrow kt_{1/2} &= \ln 2 \end{aligned}$$

$$\boxed{t_{1/2} = \frac{\ln 2}{k} \qquad \text{(first-order reaction)}}$$

Half-Life for a Second-Order Reaction

For a second-order reaction of the form A → Products:

$$\frac{1}{[A]} = kt + \frac{1}{[A_0]}$$

When $t = t_{1/2}$, $[A] = [A_0]/2$

$$\Rightarrow \frac{2}{[A_0]} = kt_{1/2} + \frac{1}{[A_0]}$$

$$\Rightarrow kt_{1/2} = \frac{1}{[A_0]}$$

$$\boxed{t_{1/2} = \frac{1}{k[A_0]} \qquad \text{(second-order reaction)}}$$

EXAMPLES

> ***Example No. 1:*** (a) Compound A undergoes a decomposition reaction, which obeys first-order kinetics, with a specific rate constant $k = 5.18 \times 10^{-4}\ s^{-1}$. Determine the concentration of A that remains 780 s after it commences to decompose at 83 °C, if the initial concentration of A is 53 mM.
> (b) How long will it take the concentration of A to decrease from 53 mM to 23 mM at 83 °C?

(a) First-order kinetics: $\ln[A] = -kt + \ln[A_0]$
Given: $k = 5.18 \times 10^{-4}\ s^{-1}$, $t = 780$ s and $[A_0] = 53$ mM ⇒ unknown = [A]
$\ln[A] = -(5.18 \times 10^{-4})(780) + \ln(53) = 3.5663$ and $[A] = 35.38$ mM

(b) First-order kinetics: $\ln[A] = -kt + \ln[A_0]$
Given: $k = 5.18 \times 10^{-4}\ s^{-1}$, $[A_0] = 53$ mM and $[A] = 23$ mM ⇒ unknown = t

$-kt = \ln[A] - \ln[A_0]$ and $kt = \ln[A_0] - \ln[A] = \ln([A_0]/[A])$, since $\log(A/B) = \log A - \log B$.
$\Rightarrow t = \ln(53/23)/(5.18 \times 10^{-4}) = 1611.58$ s = 26.9 min.

> ***Example No. 2:*** Given that the initial concentration of a compound A, being consumed in a reaction obeying first-order kinetics, is 13.7 mM, determine how long it would take the reaction to be 80% complete, given that k, the specific rate constant, is 8.3×10^{-4} s^{-1}. What is the half-life of the reaction?

First-order kinetics: $\ln[A] = -kt + \ln[A_0]$
Given: $k = 8.3 \times 10^{-4}$ s^{-1}, $[A_0] = 13.7$ mM.
Since the reaction is 80% complete, this means only 20% of the initial concentration of A remains at this stage of the reaction.

$$\Rightarrow [A] = (20/100)[A_0] = 0.2[A_0] \Rightarrow \text{unknown} = t$$

$-kt = \ln[A] - \ln[A_0]$ and $kt = \ln[A_0] - \ln[A] = \ln\{[A_0]/[A]\}$, since $\log(A/B) = \log A - \log B$. Therefore, $kt = \ln\{[A_0]/(0.2[A_0])\} = \ln 5$.

$$\Rightarrow t = (\ln 5)/(8.3 \times 10^{-4}) = 1939.1 \text{ s} = 32.3 \text{ min.}$$

$$t_{1/2} = (\ln 2)/k = (\ln 2)/(8.3 \times 10^{-4}) = 835.12 \text{ s} = 13.9 \text{ min.}$$

DETERMINATION OF THE ORDER OF A REACTION

The order of a reaction may be determined by a number of different methods. Two of the most common methods are:

Integrated Rate Equations

In this method, a reaction order is assumed, and a graph is plotted for the corresponding rate equation. This is repeated until the order yielding the best fit line is obtained.

Advantage:

1. Needs only a single kinetics experiment.

Disadvantages:

1. Assumes form of rate law and tests the assumption.
2. Depends on the accuracy of the measurement.
3. Can be sensitive to side-reactions, impurities or products.

Initial Rates

This method is much more common. The rate at the beginning of a reaction is measured and the reactant concentrations are known most accurately at this time.

Advantages:
1. No assumption is made about the form of the rate equation.
2. Accurate known concentrations are used.
3. Minimal interference by the presence of impurities or products.

Disadvantages:
1. Needs a number of separate experiments.
2. An extrapolation method is required, *i.e.* rates are measured by drawing tangential lines.

WORKING METHOD FOR THE DETERMINATION OF THE ORDER OF A REACTION BY THE METHOD OF INITIAL RATES

1. Write down the chemical equation.
2. Determine the appropriate rate equation, taking care to include the stoichiometry factors, ν_A, ν_B, ν_C, ν_D, *etc.*
3. Using the given data, determine each of the following ratios:

   $$\frac{\text{Rate 1}}{\text{Rate 2}}, \quad \frac{\text{Rate 2}}{\text{Rate 3}}, \quad \frac{\text{Rate 3}}{\text{Rate 4}}, \quad etc.$$

4. From each ratio obtained in step 3, evaluate the partial orders, x, y, z *etc.* for the reaction, remembering $x^0 = 1$, from the rules of indices and the rules of logs:
   $\log(AB) = \log A + \log B$; $\log(A/B) = \log A - \log B$; $\log A^p = p \log A$
5. Evaluate the overall order of the reaction, from the sum of the partial orders: $x + y + z + \ldots etc.$
6. Determine k, the specific rate constant for the reaction for each set of data. Find the sum of these values, and divide to get an average value.
7. Determine the correct unit of the rate constant, k.
8. Answer any riders to the question.

EXAMPLES

Example No. 1: Determine the partial orders, the overall order and the approximate value of k, the specific rate constant for the reaction: A + B + C → Products, from the following experimental data, obtained by the method of initial rates:

	[A]/M	[B]/M	[C]/M	Initial Rate/M s^{-1}
Rate 1	0.1	0.2	0.3	4.8
Rate 2	0.2	0.2	0.3	9.6
Rate 3	0.1	0.1	0.3	0.6
Rate 4	0.2	0.2	0.15	19.2

Solution:

1. A + B + C → Products
2. Rate $= -\frac{1}{1}\frac{d[A]}{dt} = -\frac{1}{1}\frac{d[B]}{dt} = -\frac{1}{1}\frac{d[C]}{dt} = k[A]^x[B]^y[C]^z$, where x, y and z are the unknown partial orders of the reaction.
3. and 4. (See note at end of question).

$$\begin{aligned}\text{Rate 1/Rate 2} &= \{k(0.1)^x(0.2)^y(0.3)^z\}/\{k(0.2)^x(0.2)^y(0.3)^z\} \\ &= (0.1)^x/(0.2)^x \quad = (0.5)^x \\ &= (4.8)/(9.6) \quad = (0.5)\end{aligned}$$

Therefore, from the rules of indices, $x = 1$.
Using $x = 1$,

$$\begin{aligned}\text{Rate 2/Rate 3} &= \{k(0.2)^1(0.2)^y(0.3)^z\}/\{k(0.1)^1(0.1)^y(0.3)^z\} \\ &= 2(2)^y = (9.6)/(0.6) = 16\end{aligned}$$

Hence, $2^y = 8$, *i.e.* $y = 3$ by inspection
(or take logs on both sides of the equation: $\ln 2^y = \ln 8 \Rightarrow y \ln 2 = \ln 8$, since $\log A^p = p \log A$, *i.e.* $y = 3$)
Using, $x = 1$ and $y = 3$,

$$\begin{aligned}\text{Rate 3/Rate 4} &= \{k(0.1)^1(0.1)^3(0.3)^z\}/\{k(0.2)^1(0.2)^3(0.15)^z\} \\ &= (0.5)(0.5)^3(2)^z \quad = \quad (0.5)^4(2)^z \\ &= (0.6)/(19.2) \quad = \quad 0.03125\end{aligned}$$

Hence, $2^z = (0.5)$, $\ln 2^z = \ln 0.5$, $z \ln 2 = \ln 0.5$, and $z = -1$.

5. Overall order of the reaction $= x + y + z = 1 + 3 - 1 = 3°$.
6. Determination of k: Rate $= k[A]^1[B]^3[C]^{-1}$.
   There are four initial rates given in the question. k must be evaluated for each of these rates and then an average value should be obtained:

   $k = (4.8)/(0.1)^1(0.2)^3(0.3)^{-1} = 1.8 \times 10^3$
   $k = (9.6)/(0.2)^1(0.2)^3(0.3)^{-1} = 1.8 \times 10^3$
   $k = (0.6)/(0.1)^1(0.1)^3(0.3)^{-1} = 1.8 \times 10^3$
   $k = (19.2)/(0.2)^1(0.2)^3(0.15)^{-1} = 1.8 \times 10^3$
   Average value of $k = 1.8 \times 10^3$

7. Units of k: k = rate/[concentration]^3 $\Rightarrow$ k has units of $Ms^{-1}/M^3 = M^{-2}\ s^{-1}$.

Answer: The reaction is first-order w.r.t. A, third-order w.r.t. B and has order −1 w.r.t. C. The overall order of the reaction is third-order. Average value of k, the specific rate constant = 1.8 × 10³ M⁻² s⁻¹.

Note: Sometimes, the partial orders can be obtained directly by inspection. For example, in the above problem, if the concentrations of B and C (0.2 and 0.3 M) are kept constant, doubling the concentration of A (0.1 to 0.2 M) doubles the rate (4.8 to 9.6 M s^{-1}). Therefore the reaction must be first-order with respect to A. However, caution must be applied here, as this skill only comes with experience. In the next example, such an initial deduction is much more difficult, and hence the working method might be a better place to begin, to evaluate the partial orders.

Example No. 2: Determine the approximate overall order for the reaction: A + B + C → Products, from the following experimental data, obtained by the method of initial rates:

	[A]/M	[B]/M	[C]/M	Initial Rate/M s^{-1}
Rate 1	0.856	0.198	0.699	2.94×10^{-4}
Rate 2	0.593	0.198	0.699	9.76×10^{-5}
Rate 3	0.391	0.699	0.699	6.91×10^{-5}
Rate 4	0.496	0.325	0.511	1.37×10^{-4}

Solution:

1. A + B + C → Products
2. Rate $= -\frac{1}{1}\frac{d[A]}{dt} = -\frac{1}{1}\frac{d[B]}{dt} = -\frac{1}{1}\frac{d[C]}{dt} = k[A]^x[B]^y[C]^z$, where x, y and z are the unknown partial orders of the reaction.

3 and 4

Rate 1/Rate 2 $= \{k(0.856)^x(0.198)^y(0.699)^z\}/\{k(0.593)^x(0.198)^y(0.699)^z\}$
$= (0.856/0.593)^x = (2.94 \times 10^{-4}/9.76 \times 10^{-5})$
$\Rightarrow (1.4435076)^x = 3.012295.$
Then, $\ln(1.4435076)^x = \ln(3.012295)$ and $x(\ln 1.4435076) = \ln 3.012295 \Rightarrow x = 3$
Rate 2/Rate 3
$= \{k(0.593)^3(0.198)^y(0.699)^z\}/\{k(0.391)^3(0.699)^y(0.699)^z\}$
$= 3.4884612 \times (0.2832618)^y = (9.76 \times 10^{-5}/6.91 \times 10^{-5}) = 1.4124457$
$(0.2832618)^y = 0.4048908.$
Then, $\ln(0.2832618)^y = \ln(0.4048908)$ and $y \ln(0.2832618) = \ln(0.4048908) \Rightarrow y = 0.72$
Rate 3/Rate 4 $=$
$\{k(0.391)^3(0.699)^{0.72}(0.699)^z\}/\{k(0.496)^3(0.325)^{0.72}(0.511)^z\}$
$= 0.8502622 \times (1.3679061)^z = (6.91 \times 10^{-5}/1.37 \times 10^{-4}) = 0.5043796 \Rightarrow (1.3679061)^z = 0.59320487.$
Then, $\ln(1.3679061)^z = \ln(0.5932048) \Rightarrow z = -1.67$

5. Overall order of the reaction $= x + y + z = 3 + 0.72 - 1.67 = 2.05$, *i.e.* approximately 2.

Answer: The approximate overall order of the reaction is second-order.

Note: Fractional orders normally imply that the mechanism of a reaction is more complicated then expected, *i.e.* consecutive reactions may be occurring.

HOW CONCENTRATION DEPENDENCE, *i.e.* THE ORDER OF A REACTION, CAN BE USED TO DEVELOP A MECHANISM FOR A REACTION

The bromination of propanone (acetone), in the presence of acid H^+, is a second-order reaction, whose rate was found to be independent of the concentration of the bromine Br_2, *i.e.*

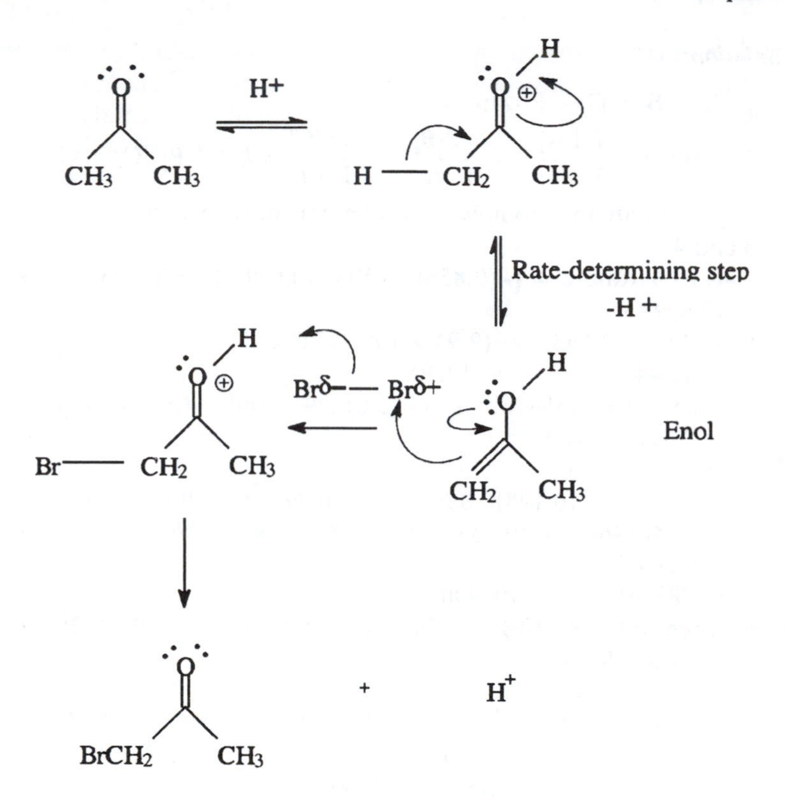

Figure 8.4 *Mechanism of the acid-catalysed bromination of propanone*

$$CH_3COCH_3 + Br_2 \xrightarrow{H^+} CH_2BrCOCH_3 + HBr$$

$$-\frac{d[\text{Propanone}]}{dt} = k[CH_3COCH_3]^1[H^+]^1[Br_2]^0$$

where propanone is CH_3COCH_3.

It was found experimentally that the bromine concentration had no effect on the rate. This means bromine cannot be involved in the ***rate-determining*** step of the reaction *i.e.* the ***slow step*** of the reaction. To explain this, a mechanism was postulated, which involved rearrangement of propanone prior to reaction (Figure 8.4).

The first step of the reaction is the loss of one of the acidic protons in propanone α to the carbonyl group, to yield an enol as the intermediate (alkene alcohol). The electron pair of the double bond of the enol then attacks the positively polarised bromine atom, to give an intermediate cation. The final step of the reaction is the loss of the proton to give the α-brominated product. The proton is regenerated at the end of the reaction, and therefore is catalytic. The enolic equilibrium concentration in propanone at 298 K is only 0.0025%. The conversion of the ketone to the enol intermediate is the ***rate-determining step***, or ***slow step*** of the reaction, since the subsequent bromination of the double bond is quite rapid. The existence of the enol intermediate and the mechanism of the reaction was first proposed, on the basis of the unique kinetic experimental results.

SUMMARY

In Chapter 8, the basic laws of kinetics have been introduced, including both zero-order and first-order reactions. A similar analysis can be proposed for second-order reactions, where the integration is slightly more difficult. In the final chapter of this text, Chapter 9, the Arrhenius Equation and a general working method to solve graphical problems will be introduced.

Chapter 9

Chemical Kinetics II: The Arrhenius Equation and Graphical Problems

MOLECULARITY

In Chapter 8, the ***order*** of a reaction was defined as the sum of the exponents of the concentration terms in the rate equation. This order is an experimentally determined quantity not to be confused with a term called ***molecularity***. Most reactions consist of a number of steps, and each individual step is known as an ***elementary reaction***. Each elementary reaction can be described by the ***molecularity*** of the process:

(a) When a single particle is the only reactant, the reaction is ***unimolecular***, *i.e.* molecularity = 1.
(b) When two particles collide, the reaction is ***bimolecular***, *i.e.* molecularity = 2.
(c) When three particles collide, the reaction is ***termolecular***, *i.e.* molecularity = 3, *etc.* However, it should be borne in mind that a full reaction may have substeps, and termolecular collisions are infrequent.

This is summarised in Table 9.1 below.

Table 9.1 *Molecularity.*

1. A → Products	***Unimolecular***	Molecularity = 1
2. A + A → Products	***Bimolecular***	Molecularity = 2
3. A + B → Products	***Bimolecular***	Molecularity = 2
4. A + B + C → Products	***Termolecular***	Molecularity = 3

Therefore, the reaction $2ClO_{(g)} \rightarrow O_{2(g)} + Cl_{2(g)}$ is an example of a bimolecular reaction, *i.e.* A + A → Products, as the two ClO gaseous molecules collide with one another to form oxygen and chlorine gases respectively as the products. In order for gaseous molecules to react, they must collide with one another. The calculation of kinetic rates from this is called ***The Collision Theory***.

THE ACTIVATION ENERGY OF A REACTION

In Chapter 1, kinetic energy was defined as the energy a body possesses by virtue of its motion, *e.g.* a moving car does work if it collides with a wall. In the Collision Theory, a reaction is said to occur when the two reacting molecules collide with a certain minimum amount of kinetic energy, *i.e.* if the energy is less than the ***activation energy*** of the reaction, E_{act}, (the minimum energy needed for a reaction to occur), the molecules rebound off one another, as shown in Figure 9.1 (a).

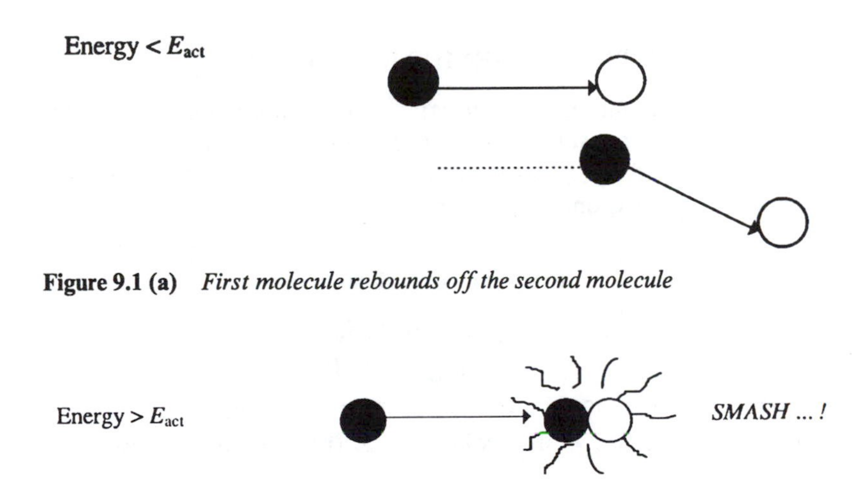

Figure 9.1 (a) *First molecule rebounds off the second molecule*

Figure 9.1 (b) *First molecule interacts with the second molecule*

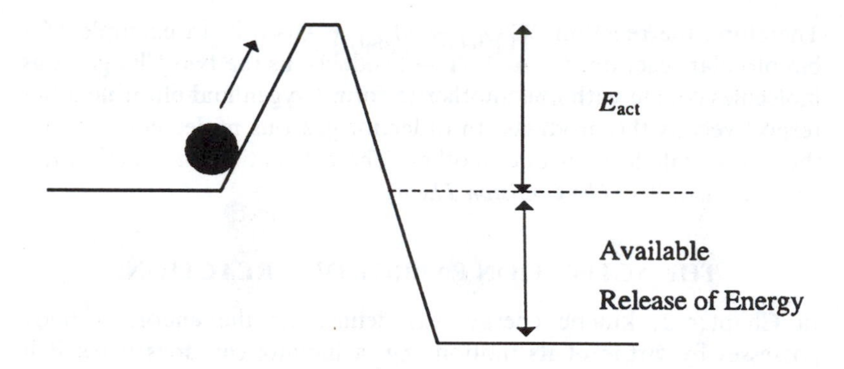

Figure 9.1 (c) *Activation energy of a reaction—minimum energy needed for a reaction to occur. Notice the initial energy barrier that must be overcome, before the reaction can take place*

The molecules behave like billiard balls. They might not collide, but if they do collide, the reaction may be small or large [Figure 9.1 (b)]. The unit of E_{act} [Figure 9.1 (c)] is the J mol^{-1}, normally expressed in kJ mol^{-1}.

THE ARRHENIUS EQUATION

The Arrhenius equation, $k = Ae^{-E_{act}/RT}$ is an expression that relates k, the specific rate constant of a reaction, to E_{act}, the activation energy of a reaction, and to the temperature, T. A is termed the Arrhenius parameter or pre-exponential factor.

Frequency of Collision

Consider the bimolecular gas-phase reaction:

$$A_{2(g)} + B_{2(g)} \rightarrow 2AB$$

In order for the reaction to occur, $A_{2(g)}$ must collide with $B_{2(g)}$.

$\Rightarrow$ Rate $\propto$ frequency of the collisions, Z (number of collisions per second)

$\Rightarrow$ Rate $= c_oZ$, where c_o is a constant of proportionality.

But, Z itself is proportional to the concentration of both A_2 and B_2.

$$i.e.\ Z \propto [A_2] \text{ and } Z \propto [B_2]$$

$Z \propto [A_2][B_2] = Z_o[A_2][B_2]$, where Z_o is another constant of proportionality.

$$\Rightarrow \qquad \boxed{\text{Rate} = c_oZ_o[A_2][B_2]}$$

The Steric Factor, *p*

For a reaction to occur, the relative orientation of the molecules at the point of collision (Figure 9.2) must be favourable.

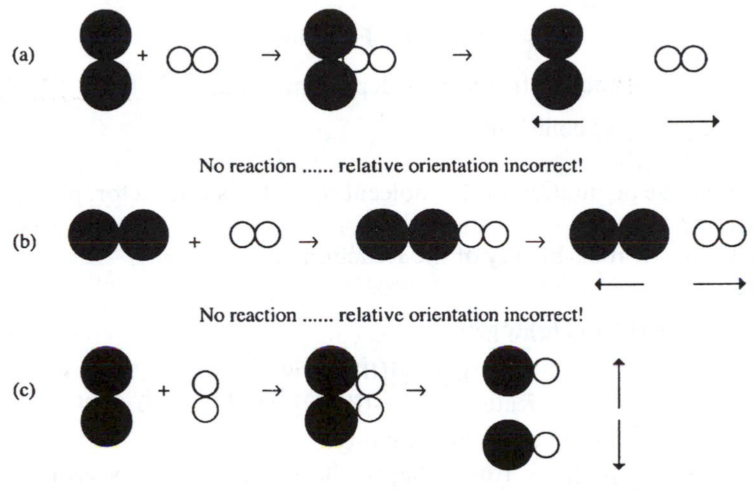

Figure 9.2 *Collisions leading to probable reaction:* $A_{2(g)} + B_{2(g)} \rightarrow 2AB$

The steric factor, *p*, is the fraction of collisions in which the molecules have a favourable relative orientation for the reaction to occur. *p* is sometimes called the probability factor, the orientation factor or the fudge factor.

$$\text{Rate} \propto p$$

Activation Energy of a Reaction, E_{act}

In any group of reactant molecules, only a fraction of molecules have energies at least equal to E_{act}, the activation energy of the reaction. It has been shown that this fraction of molecules is given by the expression:

$$e^{-E_{act}/RT}$$

where E_{act} = activation energy of the reaction, measured in J mol^{-1};

R = Universal Gas Constant = 8.314 J K^{-1} mol^{-1}; T = absolute temperature, measured in K; e = exponential = 2.71828 ....

$$\boxed{\text{Rate} \propto e^{-E_{act}/RT}}$$

Arrhenius Equation

The rate of a reaction therefore is dependent on three factors:

(a) Frequency of collision:
Rate $\propto [A_2][B_2]$
(b) Relative orientation of the molecules, *i.e.* the steric factor, p:
Rate $\propto p$
(c) The Activation Energy of the Reaction
Rate $\propto e^{-E_{act}/RT}$

Taking (a), (b) and (c) together,

$$\text{Rate} \propto p[A_2][B_2]e^{-E_{act}/RT}$$
$$\text{Rate} = c\,p[A_2][B_2]e^{-E_{act}/RT}$$

where c is a constant of proportionality.
But rate = $k[A_2][B_2]$, from Chapter 8, where k = the specific rate constant.

$$\Rightarrow \quad k[A_2][B_2] = c\,p[A_2][B_2]e^{-E_{act}/RT}$$
$$\Rightarrow \quad k = c\,pe^{-E_{act}/RT}$$
$$\Rightarrow \quad k = Ae^{-E_{act}/RT}$$

i.e. k is independent of the concentration terms. A is known as the ***frequency factor***, the ***Arrhenius parameter*** or the ***pre-exponential*** term. A has the same units as k, *i.e.* zero-order reaction, Ms^{-1}; first-order reaction, s^{-1}, *etc.*, where the order of a reaction is the sum of the exponents of the concentration terms in the rate equation, as described in Chapter 8.
If natural logs are now taken on both sides of this equation:

$$\ln k = \ln Ae^{-E_{act}/RT}$$
$$\ln k = \ln e^{-E_{act}/RT} + \ln A \quad (\text{since } \log AB = \log A + \log B)$$
$$\ln k = (-E_{act}/RT)\ln e + \ln A \quad (\text{since } \log A^x = x \log A)$$
$$\Rightarrow \ln k = (-E_{act}/RT) + \ln A \quad (\text{since } \ln e = 1)$$

This generates a much more useful expression for the Arrhenius equation, as this represents the equation of a line, *i.e.* if a graph (Figure 9.3) is plotted of $\ln k$ *versus* $1/T$, the slope or gradient of the graph, $m = \Delta y/\Delta x$, is given by $-E_{act}/R$, from which E_{act}, the activation energy of the reaction, can be determined. The pre-exponential factor, or Arrhenius parameter A can be derived from the intercept, c, of the graph, *i.e.* $c = \ln A$.

$$\ln k = -\frac{E_{act}}{R}\frac{1}{T} + \ln A \quad \text{Arrhenius equation}$$

$$y = m \quad x + c$$

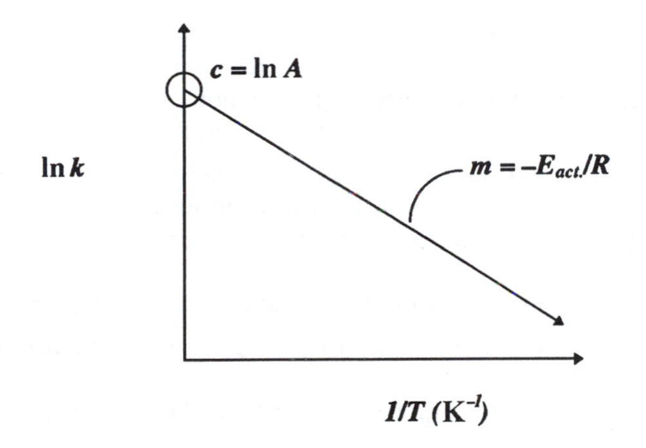

Figure 9.3 *Plot of ln k versus 1/T for the Arrhenius equation*

WORKING METHOD FOR THE SOLUTION OF GRAPHICAL PROBLEMS IN KINETICS

1. Read the question carefully.
2. Identify the tabulated data given—tables of data suggest that a graph needs to be plotted. Remember, this may not be specified in the problem.
3. Convert all units to the SI system, *i.e.* T should be expressed in K and E_{act} in J mol^{-1} (or kJ mol^{-1}).
4. Having identified the data, try to establish a linear correlation between the two sets of data—remember it is *not* simply a case of plotting one set of data points on the x-axis and one set on the y-axis. Identify the linear equation in question:

i.e. zero-order $[A] = -kt + [A_0]$
$y = mx + c$
first-order $\ln[A] = -kt + \ln[A_0]$
$y = mx + c$
second-order $1/[A] = kt + 1/[A_0]$
$y = mx + c$
for A → Products
Arrhenius $\ln k = -E_{act}/RT + \ln A$
$y = mx + c$

Remember if $t_{1/2}$ values are given, this is equivalent to a set of k values since:

(a) $t_{1/2} = [A_0]/(2k)$ for a zero-order reaction
(b) $t_{1/2} = (\ln 2)/k$ for a first-order reaction
(c) $t_{1/2} = 1/(k[A_0])$ for a second-order reaction

as shown in Chapter 8. Also, it is possible that a table of pressures may be given instead of concentration values [A]. A temperature T usually indicates that the Arrhenius expression is involved.

5. Create a table of the appropriate data, taking special care of:

(a) logs—did you use natural logs, $\log_e = \ln$, to the base e, for example?
(b) did you convert direct values (T) to their reciprocal values ($1/T$)?
(c) units (*e.g.* logs: dimensionless; $1/T$: K^{-1}, *etc.*)

Add as many extra columns as required. Keep the tabulated data vertical (*i.e.* go down the page); this will ensure you do not run out of space!

x-axis/unit	*y*-axis/unit

6. Examine the table in step 5. From this, write down the maximum and minimum values of x and y respectively:

   Maximum value of x = ;   Minimum value of x = ;
   Maximum value of y = ;   Minimum value of y = .

   This now determines an appropriate scale for the graph. At this point, you might want to return to step 5, and 'round off' any numbers for plotting purposes, but be careful with the number of ***significant figures*** for accuracy. Add an additional column in the table if necessary.
7. Draw the graph on graph paper, and remember the following points.

   (a) Every graph must have a title.
   (b) Label the two axes.
   (c) Include the units on the axes, but remember, there are no units for logarithmic values.
   (d) Maximise the scale of the graph for accuracy.
   (e) Draw the best-fit line through the set of points (Figure 9.4). It is not essential that the line contains any of these experimental points.

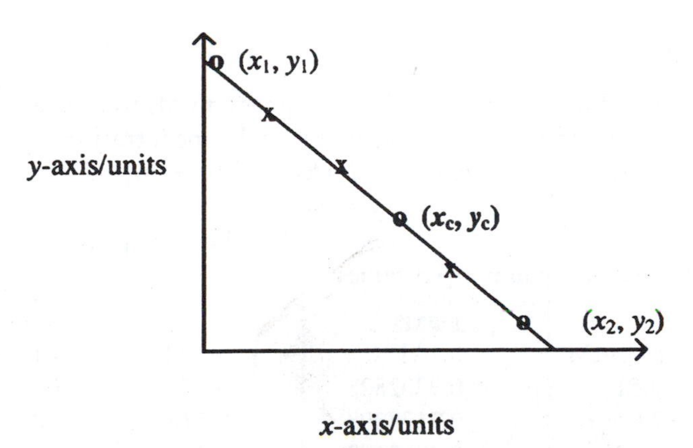

Figure 9.4 *Plot of y versus x*

8. Determine the slope or gradient of the graph, by choosing two independent points on the line at the two extremities, (x_1, y_1) and (x_2, y_2) respectively.

Then: $m = \Delta y/\Delta x = (y_2 - y_1)/(x_2 - x_1)$, and do not forget that the slope has units too.

9. The intercept of the graph, c, is then determined by examining where the graph cuts the y-axis at $x = 0$. The units of c are obviously the y-axis units. If, however, you find from the scale of your graph that $x = 0$ is not included, c can still be determined, without extrapolating (extending) the graph. To do this, choose another independent point (x_c, y_c) on the line in the centre of the graph, and use the formula:
   $y = mx + c \Rightarrow y_c = mx_c + c \Rightarrow c = y_c - mx_c$, since m has already been determined in step 8.
10. From the values of m and c, determine the unknown parameter(s), *e.g.* E_{act}, A, *etc.*
11. Answer any riders to the question.

WORKED EXAMPLES

Example No. 1: For the reaction A → Products, the following data were obtained at 350 °C. Show graphically that this is a first-order reaction and determine k, the specific rate constant for the reaction:

t/min	0	10	20	30	40
[A]/mM	2.51	2.05	1.53	1.10	0.86

Solution:

1. Read the question carefully—a graph needs to be drawn!
2. Two sets of data are given: time and concentration.
3. If first-order kinetics ⇒ $\ln[A] = -kt + \ln[A_0]$
   $y = mx + c$
   $m = -k$ and $c = \ln[A_0]$
4. Need to obtain ln[A] values:

	***y*-axis**		***x*-axis**
[A]/mM	ln[A]	ln[A]	t/min
2.51	0.9202828	0.920	0
2.05	0.7178398	0.718	10
1.53	0.4252677	0.425	20
1.10	0.0953102	0.095	30
0.86	−0.1508229	−0.151	40

5. Maximum value of $x = 40$; minimum value of $x = 0$.
   ⇒ A suitable x-axis scale is 0 to 50.

Maximum value of $y = 0.92$; minimum value of $y = -0.151$.
$\Rightarrow$ A suitable y-axis scale is -0.6 to 1.0.

6. Draw the graph: this is represented in Figure 9.5.
   Straight line graph: therefore first-order reaction confirmed.
7. Slope of graph: $(x_1, y_1) = (0, 0.96)$; $(x_2, y_2) = (50, -0.44)$
   $m = \Delta y/\Delta x = (y_2 - y_1)/(x_2 - x_1) = (-0.44 - 0.96)/(50 - 0)$
   $= -0.028\ \text{min.}^{-1}$
   $m = -0.028\ \text{min}^{-1}$; then, as $m = -k$, $k = 0.028$ min.

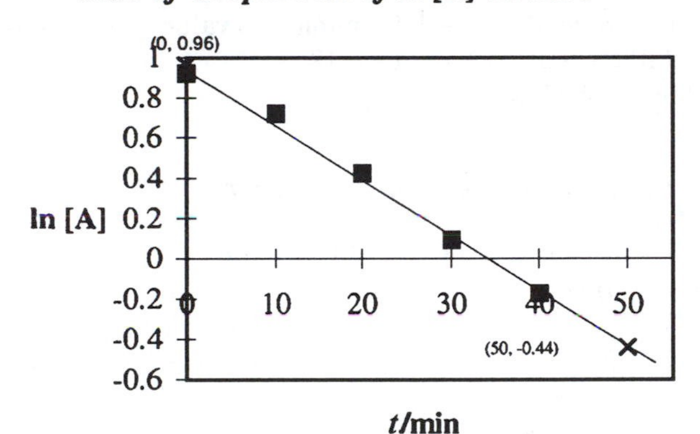

Figure 9.5 *Plot of ln [A] versus t for example 1*

Example No. 2: Find the activation energy, E_{act}, and the pre-exponential factor, A, for the reaction $A_{2(g)} + B_{2(g)} \rightarrow 2AB$, given that $R = 8.314\ \text{J K}^{-1}\ \text{mol}^{-1}$ and the following data:

T/K	538	637	734	792	835
$k/\text{M}^{-1}\text{s}^{-1}$	7.19×10^{-7}	6.32×10^{-5}	2.37×10^{-3}	7.65×10^{-2}	3.29×10^{-1}

Solution:

1. Read the question carefully—a graph needs to be drawn!
2. Two sets of data are given: T, the temperature and k, the specific rate constant. Temperature suggests that the expression for the Arrhenius equation is involved.
3. Arrhenius expression: $\ln k = -E_{act}/RT + \ln A$
4. Need to obtain $\ln k$ and $1/T$ values:

	y-axis		**x-axis**
k	$\ln k$	T	$1/T$
$M^{-1}s^{-1}$		K	$K^{-1} \times 10^{-3}$
7.19×10^{-7}	-14.15	538	1.86
6.32×10^{-5}	-9.67	637	1.57
2.37×10^{-3}	-6.04	734	1.36
7.65×10^{-2}	-2.57	792	1.26
3.29×10^{-1}	-1.11	835	1.20

5. Maximum value of $x = 1.86$; minimum value of $x = 1.20$.
   $\Rightarrow$ A suitable x-axis scale is 1.0 to 2.0.
   Maximum value of $y = -1.11$; minimum value of $y = -14.15$.
   $\Rightarrow$A suitable y-axis scale is 2 to -18.
6. Draw the graph (Figure 9.6).

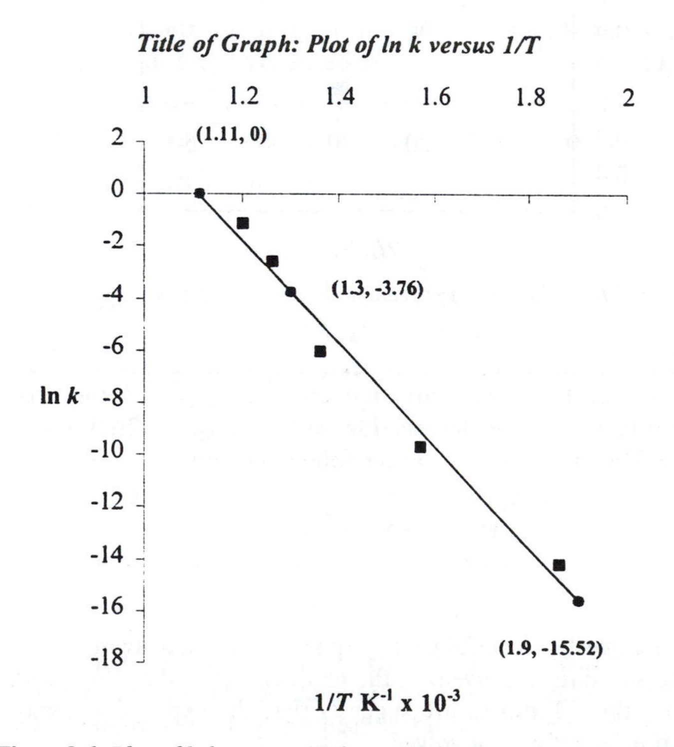

Figure 9.6 *Plot of ln k versus 1/T for example 2*

7. Slope of graph: $(x_1, y_1) = (1.11, 0)$; $(x_2, y_2) = (1.90, -15.52)$
$m = \Delta y/\Delta x = (y_2 - y_1)/(x_2 - x_1) = (-15.52 - 0)/(1.90 - 1.11) = -19.646 \times 10^3$ K.
$m = -E_{act}/R = -E_{act}/(8.314) = -18.354 \times 10^3 \text{ K} \Rightarrow E_{act} \approx 163.34 \text{ kJ mol}^{-1}$.
8. Intercept of the graph: $y = mx + c \Rightarrow c = y_c - mx_c$. Choosing $(x_c, y_c) = (1.3, -3.76)$
$\Rightarrow c = (-3.76) - (-19.646)(1.3) = 21.78 = \ln A$
$\Rightarrow A = 2.88 \times 10^9 \text{ M}^{-1}\text{s}^{-1}$.

RATE CONSTANTS AND TEMPERATURE

The Arrhenius equation states that:

$$\ln k = -E_{act}/RT + \ln A$$

If k_1 is the rate constant at temperature T_1, and k_2 is the rate constant at temperature T_2, an expression can be derived relating these four parameters:

$$\ln k_1 = \frac{-E_{act}}{RT_1} + \ln A$$

$$\ln k_2 = \frac{-E_{act}}{RT_2} + \ln A$$

$$\ln k_1 - \ln k_2 = (-E_{act}/RT_1) + \ln A + (E_{act}/RT_2) - \ln A$$
$$= (E_{act}/R)(1/T_2 - 1/T_1)$$

But since $\ln k_1 - \ln k_2 = \log \text{A} - \log \text{B} = \log (\text{A/B})$, then:

$$\ln (k_1/k_2) = (E_{act}/R)(T_1 - T_2)/(T_1T_2)$$

$$\ln(k_1/k_2) = \frac{E_{act}}{R}\,\frac{(T_1 - T_2)}{(T_1)(T_2)}$$

$$y = m\,x + c$$

i.e. if a graph is drawn of $\ln (k_1/k_2)$ *versus* $(T_1 - T_2)/(T_1T_2)$, a straight line of positive gradient $m = (E_{act}/R)$ would be obtained, from which the activation energy E_{act} could subsequently be determined. The graph would pass through the origin, (0,0), as the intercept $c = 0$. The major value of the above expression is that given any four of the five variables, k_1, k_2, T_1, T_2 or E_{act}, the fourth can be determined from the equation.

Examples

> ***Example No. 1:*** Determine the rate constant at 42 °C for the hydrolysis of a sugar, S, given that the activation energy of the reaction is 132 kJ mol^{-1} and the rate constant at 53 °C is 1.12×10^{-3} dm^3 mol^{-1} s^{-1} and $R = 8.314$ J K^{-1} mol^{-1}.

1. Identify the data in the question:

   T_1 = 42 °C
   T_2 = 53 °C
   k_2 = 1.12×10^{-3} dm^3 mol^{-1} s^{-1}
   E_{act} = 132 kJ mol^{-1}
   R = 8.314 J K^{-1} mol^{-1}

2. Convert all units to SI:

   T_1 = (42 + 273) K = 315 K
   T_2 = (53 + 273) K = 326 K
   k_2 = 1.12×10^{-3} dm^3 mol^{-1} s^{-1}
   E_{act} = 132 kJ mol^{-1} = 132 000 J mol^{-1}
   R = 8.314 J K^{-1} mol^{-1}

3. Identify the unknown(s) in the question to be determined: the rate constant k_1 at temperature T_1.
4. What expression is involved?

   $$\ln (k_1/k_2) = [(E_{act}/R)(T_1 - T_2)]/(T_1 T_2)$$

5. Rearrange the equation for k_1 *before* substituting the numerical values:

   $$\ln k_1 - \ln k_2 = [(E_{act}/R)(T_1 - T_2)]/(T_1 T_2)$$
   $$\ln k_1 = [(E_{act}/R)(T_1 - T_2)]/(T_1 T_2) + \ln k_2$$

6. Substitute the values into the equation, taking care of signs and brackets:

   $$\ln k_1 = (132\,000/8.314)(315-326)/[(315)(326)] + \ln (1.12 \times 10^{-3})$$
   $$= -8.4951295$$
   $$k_1 = e^{-8.4951295} = 2.045 \times 10^{-4}$$

7. What are the units of k_1? $k_1 = 2.045 \times 10^{-4}$ dm^3 mol^{-1} s^{-1}

   Answer: $\boldsymbol{k_1 = 2.045 \times 10^{-4}}$ ***dm³ mol⁻¹ s⁻¹***

Example No. 2: A first-order reaction, A → B, takes 5.8 minutes at 32 °C to complete a 15% loss of A. If the activation energy of the reaction is 65.2 kJ mol^{-1}, determine the half-life of the reaction at 45 °C, given that R = 8.314 J K^{-1} mol^{-1}.

1. Identify the data in the question:

$$\begin{aligned} T_1 &= 32\,°\text{C} \\ t_1 &= 5.8\text{ min} \\ T_2 &= 45\,°\text{C} \\ E_{act} &= 65.2\text{ kJ mol}^{-1} \\ R &= 8.314\text{ J K}^{-1}\text{ mol}^{-1} \end{aligned}$$

2. Convert all units to SI:

$$\begin{aligned} T_1 &= (32 + 273)\text{ K} &&= 305\text{ K} \\ t_1 &= 5.8\text{ min} &&= 348\text{ s} \\ T_2 &= (45 + 273)\text{ K} &&= 318\text{ K} \\ E_{act} &= 65\,200\text{ J mol}^{-1} \\ R &= 8.314\text{ J K}^{-1}\text{ mol}^{-1} \end{aligned}$$

3. Identify the unknown, to be evaluated: the half-life, $t_{1/2}$ at temperature 45 °C (T_2).
4. What expressions are involved?
   (a) $\ln(k_1/k_2) = (E_{act}/R)(T_1 - T_2)/(T_1T_2)$
   (b) $\ln[A] = -kt + \ln[A_0]$, for a first-order reaction.
   (c) $t_{1/2} = (\ln 2)/k$ for a first-order reaction.
5. To determine $t_{1/2}$, k_2 must first be evaluated from equation (a). Rearrange the equation for k_2 *before* substituting the numerical values:

$$\begin{aligned} \ln k_1 - \ln k_2 &= (E_{act}/R)(T_1 - T_2)/(T_1T_2) \\ \ln k_2 &= -(E_{act}/R)(T_1 - T_2)/(T_1T_2) + \ln k_1 \end{aligned}$$

6. Substitute the values into the equation, taking care of signs and brackets:

$$\begin{aligned} \ln k_2 &= -(65\,200/8.314)(305 - 318)/[(305)(318)] + \ln k_1 \\ &= 1.051124 + \ln k_1 \quad (\dagger) \end{aligned}$$

However, no explicit value is given in the question for k_1. Hence, this must be determined from another expression, before substitution into the above equation.

1° reaction: $\ln[A] = -kt + \ln[A_0]$. Rearrange this equation for k, before substituting the numerical values:

$kt = \ln[A_0] - \ln[A] \Rightarrow k = (1/t)(\ln[A_0] - \ln[A]) = (1/t)(\ln[A_0]/[A])$, (since $\log(A/B) = \log A - \log B$).

But $t_1 = 348$ s at 32 °C, when [A] = 85% [A_0].

$\Rightarrow k_1 = (1/348)(\ln[A_0]/0.85\ [A_0]) = (1/348)(\ln 1.1764706) = 0.000467\ s^{-1}$.

Therefore, from step 6 (†)

$\ln k_2 = 1.051124 + \ln k_1 = 1.051124 + \ln 0.000467 = -6.6180573$

Hence, $k_2 = e^{-6.6180573} = 0.001336023$.

7. What are the units? $k_2 = 0.001336023\ s^{-1}$
8. $t_{1/2} = (\ln 2)/(0.001336023) = 518.8$ s, as $k = (\ln 2)/t_{1/2}$, for a first-order reaction.

Answer: $t_{1/2} = 518.8$ ***s***

SUMMARY OF THE TWO CHAPTERS ON CHEMICAL KINETICS

There now follow the five sets of equations which should be remembered for kinetics type problems at this level. In the following two sections, a multiple-choice test and three longer questions, involving the working methods covered in Chapters 8 and 9, respectively, are given:

1. Zero-order $[A] = -kt + [A_0]$ for A → Products
   $y = mx + c$
   Half-Life $t_{1/2} = [A_0]/(2k)$
2. First-order $\ln[A] = -kt + \ln[A_0]$ for A → Products
   $y = mx + c$
   Half-Life $t_{1/2} = (\ln 2)/k$
3. Second-order $1/[A] = kt + 1/[A_0]$ for A → Products
   $y = mx + c$
   Half-Life $t_{1/2} = 1/(k[A_0])$
4. Arrhenius $\ln k = -E_{act}/(RT) + \ln A$
   $y = mx + c$
5. $\ln(k_1/k_2) = (E_{act}/R)[(T_1 - T_2)/(T_1T_2)]$

MULTIPLE-CHOICE TEST

1. The reaction A + B → Products obeys the rate law:

$$-\frac{d[A]}{dt} = k[A]^{-1}[B]^2$$

How many of the following statements are true?

* The reason is second-order with respect to B.
* The reaction is third-order overall.
* The reaction is first-order with respect to A.
* A and B are consumed at the same rate in the reaction.

(a) 4 (b) 3 (c) 2 (d) 1

2. If the half-life of the first-order decomposition of A at 34 °C is 2.05×10^{4} s, what is the value of the rate constant in s^{-1}?

   (a) 2.05×10^{-4} (b) 1.03×10^{4} (c) 3.38×10^{-5} (d) 3.38×10^{5}

3. The rate constant doubles when the temperature is increased from 35 to 53 °C. What is the activation energy of the reaction in kJ mol^{-1}? ($R = 8.314$ J K^{-1} mol^{-1}).

   (a) -3.2146 (b) 32146 (c) 0.594 (d) 32.146

4. A zero-order reaction A → Products commences with $[A_0] =$ 0.25 M, and after 2 minutes [A] = 0.032 M. Determine k, the rate constant for the reaction, in Ms^{-1}.

   (a) 0.109 (b) 0.0018 (c) -0.109 (d) 0.018

5. In the reaction $2A_{(g)} + B_{(g)} \rightarrow 2C_{(g)}$ the following initial rates were observed for certain reactant concentrations:

	[A]/M	[B]/M	Initial rate/ M h^{-1}
Rate 1	0.25	0.25	3.2
Rate 2	1.00	0.25	12.8
Rate 3	1.00	1.00	6.4

   What is the rate equation for the reaction?

   (a) Rate $= k[A]^{0.5}[B]^{1}$ (b) Rate $= k[A]^{1}[B]^{-0.5}$
   (c) Rate $= k[A]^{1}[B]^{0.5}$ (d) Rate $= k[A]^{0.5}[B]^{-1}$

6. For a second-order reaction, A → Products, what are the units of k, the specific rate constant?

   (a) $(time)^{-1}$ (b) $(conc.)^{-1}.(time)^{-1}$
   (c) $(conc.).(time)^{-2}$ (d) $(conc.).(time)^{-1}$

LONG QUESTIONS ON CHAPTERS 8 AND 9

1. A reaction obeys the stoichiometric equation A + 2B → 2C. Rates of formation of C at various concentrations of A and B are given in the following table:

	[A]/M	[B]/M	Initial rate/M s^{-1}
Rate 1	3.5	2.33	8.6×10^{-4}
Rate 2	1.05	7	2.87×10^{-4}
Rate 3	1.05	21	2.58×10^{-3}

Determine the approximate overall order of the reaction.

2. Evaluate the activation energy of the reaction, E_{act}, for A + 2B → C, using a graphical procedure, given the following relative rate constants at various temperatures (R = 8.314 J K^{-1} mol^{-1}).

T/°C	77	102	127	154	178	204
k	2	6	18	54	162	486

3. In a given reaction, $k = 2.6 \times 10^{-11}$ s^{-1} at 32 °C and 8.7×10^{-11} s^{-1} at 43 °C. Determine E_{act}, the activation energy of the reaction, the pre-exponential factor, A, and the value of k at 55 °C, given that R = 8.314 J K^{-1} mol^{-1}.

Answers to Problems

Chapter 1

Multiple-Choice Test (p. 15)

1.(c); 2 (b); 3.(d); 4.(a); 5.(a); 6.(d).

Chapters 2 and 3

Multiple-Choice Test (p. 34)

1.(c); 2.(b); 3.(a); 4.(d); 5.(a).

Long Questions (p. 35)

1. -1366.81 kJ mol^{-1}; 2. 158.83 kJ mol^{-1}, non-spontaneous reaction; 3. -1193.78 kJ mol^{-1}, spontaneous reaction.

Chapters 4 and 5

Multiple-Choice Test (p. 61)

1.(b); 2.(c); 3.(d); 4.(a); 5.(a); 6.(c).

Long Questions (p. 62)

1. $[H_3O]^+ = [HCO_3]^- = 2.51 \times 10^{-4}$ M; $[CO_3]^{2-} = 5.6 \times 10^{-11}$ M; 2. 4.4; 3. 9.9.

Chapters 6 and 7

Multiple-Choice Test (p. 110)

1.(c); 2.(a); 3.(b); 4.(d); 5.(d); 6.(d); 7.(a); 8.(a); 9.(c); 10.(c).

Long Questions (p. 111)

1. 218.09 kJmol^{-1}, non-spontaneous reaction; 2. 1.04 V; 3. 6.72×10^{19}, spontaneous reaction; 4. 858 cm^3; 5. 2.49 h; 6. III .

Chapters 8 and 9

Multiple-Choice Test (p. 143)
1.(c); 2.(c); 3.(d); 4.(b); 5.(b); 6.(b)

Long Questions (p. 143)
1. 4.7; 2. 59730 J K^{-1} mol^{-1}; 3. 87.984 kJ mol^{-1}, 30461.2 s^{-1}, 2.43×10^{-10} s^{-1}.

Further Reading

H.F. Holtzclaw, W.R. Robinson and J.D. Odom, 'General Chemistry', D.C. Heath and Company, 1991.
P.W. Atkins, 'The Elements of Physical Chemistry', Oxford University Press, 1996.
G.M. Barrow, 'Physical Chemistry', McGraw-Hill, New York, 1996.
P.W. Atkins, 'Physical Chemistry', Oxford University Press, 1994.
I.N. Levine, 'Physical Chemistry', McGraw-Hill, New York, 1995.

Periodic Table of the Elements

	1	2	3 (3d, 4d, 5d)	4	5	6	7	8	9	10	11	12	13	14	15	16	17	18	
1s	1 H 1·0079																	2 He 4·0026	
2s	3 Li 6·941	4 Be 9·01218											5 B 10·81	6 C 12·011	7 N 14·0067	8 O 15·9994	9 F 18·9984	10 Ne 20·179	2p
3s	11 Na 22·98977	12 Mg 24·305											13 Al 26·9815	14 Si 28·0855	15 P 30·9738	16 S 32·06	17 Cl 35·453	18 Ar 39·948	3p
4s	19 K 39·0983	20 Ca 40·08	21 Sc 44·9559	22 Ti 47·88	23 V 50·9415	24 Cr 51·996	25 Mn 54·938	26 Fe 55·847	27 Co 58·9332	28 Ni 58·69	29 Cu 63·546	30 Zn 65·38	31 Ga 69·72	32 Ge 72·59	33 As 74·9216	34 Se 78·96	35 Br 79·904	36 Kr 83·80	4p
5s	37 Rb 85·4678	38 Sr 87·62	39 Y 88·9059	40 Zr 91·22	41 Nb 92·9064	42 Mo 95·94	43 Tc (98)	44 Ru 101·07	45 Rh 102·9055	46 Pd 106·42	47 Ag 107·868	48 Cd 112·41	49 In 114·82	50 Sn 118·69	51 Sb 121·75	52 Te 127·60	53 I 126·9045	54 Xe 131·29	5p
6s	55 Cs 132·9054	56 Ba 137·33	57 ♯La 138·9055	72 Hf 178·49	73 Ta 180·9479	74 W 183·85	75 Re 186·207	76 Os 190·2	77 Ir 192·22	78 Pt 195·08	79 Au 196·9665	80 Hg 200·59	81 Tl 204·383	82 Pb 207·2	83 Bi 208·9804	84 Po (209)	85 At (210)	86 Rn (222)	6p
7s	87 Fr (223)	88 Ra 226·0254	89 *Ac 227·0278																

( ) mass numbers of most stable isotope

♯ LANTHANUM SERIES

4f	58 Ce 140·12	59 Pr 140·9077	60 Nd 144·24	61 Pm (145)	62 Sm 150·36	63 Eu 151·96	64 Gd 157·25	65 Tb 158·9254	66 Dy 162·50	67 Ho 164·9304	68 Er 167·26	69 Tm 168·9342	70 Yb 173·04	71 Lu 174·967

* ACTINIUM SERIES

5f	90 Th 232·0381	91 Pa 231·0359	92 U 238·0389	93 Np 237·0482	94 Pu (244)	95 Am (243)	96 Cm (247)	97 Bk (247)	98 Cf (251)	99 Es (252)	100 Fm (257)	101 Md (258)	102 No (259)	103 Lr (260)

Subject Index